Advances in Mushroom Science

Advances in Mushroom Science

Dr. Ram Dutta

SSPH

SATISH SERIAL PUBLISHING HOUSE
403, Express Tower, Commercial Complex
Azadpur, Delhi-110033 (India)
Telefax : 011 - 47073040
e-mail : info@satishserial.com, hkjain1975@yahoo.com
Website : www.satishserial.com

Published by :

SATISH SERIAL PUBLISHING HOUSE

403, Express Tower, Commercial Complex, Azadpur,
Delhi-110033 (India) Telefax : 011 - 47073040
E-mail : info@satishserial.com, hkjain1975@yahoo.com

Print Again, 2026

ISBN 978-93-85055-90-4

Composed, Designed & Printed in India

I dedicate this book to those who died after eating poisonous mushrooms.

Ram Dutta

Preface

Ever-increasing protein gap, an important facet of malnutrition, in developing countries has necessitated the search for alternative sources of protein especially in India where production of pulses, the traditional source of protein, has not kept pace with population growth. In past many alternate sources of protein have been tried and rejected due to unpalatability and consumer resistance. Thus, search for alternative sources of protein, for a predominantly vegetarian population of country, has to concentrate on a vegetable protein. Mushrooms have been recognised as vegetarian food rich in nutrition particularly protein. Mushrooms with their flavour, texture, nutritional value, very high productivity per unit area and time, less dependence on land and ability to grow on a variety of residual agricultural wastes, have rightly been identified as a food source to fight malnutrition in developing countries.

Mushrooms, in India can be exploited well as the climatic as well as biomass requirements are available round year. The three types of mushrooms, white button (*Agaricus bisporus*), Oyster (*Pleurotus* spp.) and Paddy straw (*Volvariella* spp.) can be grown round the year rolling one over another. A fourth mushroom, shiitake (*Lentinus edodes*) is also being exploited in India to meet the requirement of dry mushrooms.

Presently literature on mushrooms is scattered in various forms including bulletins, pamphlets and books. This book has been written to consolidate the elements of mushroom science together in a nutshell form. The book comprises 23 chapter excluding 'Glossary' of various terms and 'Appendices' for special references. It is, therefore, anticipated that this book will serve the purpose of mushroom growers, students, teachers, and researchers as well as housewives to fulfill their requirements.

I express my gratitude to the writers/owners/disclaimers of the numerous information, which has been consulted for preparing this manuscript. I am indebted to the individuals for their assistance provided to me at the time of exigencies. In spite of the best efforts, it is possible that some errors may have crept into the compilation. I will be grateful to readers if errors are pointed out so that necessary corrections are incorporated in the future editions of the book. I humbly welcome critical suggestions for further improvement of this book. I finally thank 'SATISH SERIAL PUBLISHING HOUSE' for cooperating and bringing this book in present form.

April 4, 2006 Ram Dutta

CONTENTS

Chapter 1
INTRODUCTION

Mushroom science is the discipline that is concerned with the principles and practices of mushroom cultivation. Mushrooms neither belong to the plants nor to animals. They belong to a separate group of organisms called fungi. They lack the usual green matter present in plants and grow on dead and decaying organic materials. The cultivated edible mushrooms may be defined, as *"Mushrooms are spore bearing fruiting bodies that are desirable as food in the market place and that can be cultivated in the quantity either in confined areas under precisely controlled conditions manipulated by man to enhance its production or in wider outdoor areas"*.

The earliest word in the Sanskrit for mushroom appears to be *Ksumpa,* which in present day Hindi, has evolved as *Khumbi.* The *Chhatra* is a later one, which is particularly used for the "fleshy cap" of mushrooms. Mushrooms are being called with several names including *Kukurmutta, Kavaka, Bhoomi Kavak and Bhustrna* in Hindi and allied languages.

It is believed that Aryans while invasion to India (around 1500 B.C.) carried a product of intoxicating mushroom called 'Soma'. The Soma was mostly used in the religious rites performed by the Aryans. According to Wasson (1967) Soma, based on Rig Veda, referred to *Amanita muscaria* (the fly agaric). Some others reports also confirm that the ancient Soma was extracted from the *A. muscaria.*

Mushrooms grow all over the globe with a high concentration in the Northern hemisphere and fewer in the Southern hemisphere. Usually one type of mushroom grows in a specific area and that area becomes known as a place to harvest that species. Because mushroom spores are so tiny and light, it's easy for them to be carried by winds and birds to locations not necessarily typical for that variety.

In the mid 1600's, Parisian melon farmers discovered that they could cultivate the common mushroom known today as *Agaricus* in their melon fields. Two hundred years later they learned that caves were the ideal environments because the climate was stable. Louis XIV may have been France's earliest mushroom grower. Today, mushrooms are grown in mushroom houses where the climate is completely controlled.

There are more than 38,000 species of fungi that can be considered as mushrooms with at least 2,000 species showing various degrees of edibility. Of these about 180 species from 70 genera are growing in India. In addition to these 180 species, there are other several species found in the tribal areas of Himalyan and Plain regions of India. In India, the highest production of mushroom is being achieved in the states of Punjab, Tamil Nadu, Madhya Pradesh, Andhra Pradesh, Himachal Pradesh, and Uttar Pradesh.

Mushrooms and toadstools

The term, Mushroom is said to be the conspicuous umbrella-shaped fruiting body (sporophore) of certain fungi, typically of the order Agaricales in the class Basidiomycetes and some other groups. Popularly, the term **mushroom** is used to identify the edible sporophores, whereas the term **toadstool** is often reserved for inedible or poisonous sporophores. There is, however, no scientific distinction between the two names, and either can be properly applied to any fleshy fungus fruiting structure.

Umbrella-shaped sporophores are found chiefly in the family Agaricaceae, members of which bear thin, bladelike gills on the undersurface of the pileus (Cap) from which the spores are shed. The sporophore of an agaric necessarily consists of a pileus and stipe (stalk). The sporophore emerges from an extensive underground network of threadlike mycelium called strands.

Mushroom mycelia may live hundreds of years or die in a few months, depending on the available food supply. As long as nourishment is available and temperature and moisture are suitable, a mycelium will produce a new crop of sporophores each year during its fruiting season.

Fruiting bodies of some mushrooms occur in arcs or rings called fairy rings. The mycelium starts from a spore falling in a favourable spot and producing hyphae that grow out in all directions, eventually forming a circular mat of underground hyphal threads. Fruiting bodies, produced near the edge of this mat, may widen the ring for hundreds of years.

A few of the mushrooms belong to the family Boletaceae, which bear spores in an easily detachable layer (the gills) on the underside of the cap. ***The agarics and boletes include most of the forms known as mushrooms***. Among these are the hydnums or hedgehog mushrooms, which have teeth, spines, or warts on the undersurface of the cap (*e.g., Dentinum repandum, Hydnum imbricatum*) or at the ends of branches (*e.g., Hydnum coralloides, Hericium caput-ursi*).

The polypores (shelf fungi) or bracket fungi (order Polyporales) have tubes under the cap as in the boletes, but they are not in an easily separable layer. Polypores usually grow on living or dead trees, sometimes as destructive pests. Many of them renew growth each year and thus produce annual growth layers by which their age can be estimated. Examples include the dryad's saddle (*Polyporus squamosus*), the beefsteak fungus (*Fistulina hepatica*), the sulfur fungus (*Polyporus sulphureus*), the artist's fungus (*Ganoderma applanatum*), and species of the genera *Fomes* and *Trametes*.

The clavarias, or club fungi (*e.g., Clavaria, Ramaria*), are shrub-like, club-like, or coral-like in growth habit. One club fungus, the cauliflower fungus (*Sparassis crispa*), has flattened clustered branches that lie close together, giving the appearance of the vegetable cauliflower. The cantharelloid fungi (*Cantharellus* and its relatives) are club-, cone-, or trumpet-shaped mushroom-like forms with an expanded top bearing coarsely folded ridges along the underside and descending along the stalk. Examples include the highly prized edible chanterelle (*C. cibarius*) and the horn-of-plenty mushroom (*Craterellus cornucopioides*).

Puffballs (order Lycoperdales), stinkhorns, earthstars (a kind of puffball), and bird's nest fungi are usually treated with the mushrooms. The morels (*Morchella, Verpa*) and false morels or lorchels (*Gyromitra, Helvella*) of the class Ascomycetes are popularly included with the true mushrooms because of their shape and fleshy structure. They resemble a deeply folded or pitted conelike sponge at the top of a hollow stem. Some are among the most highly prized edible fungi (*e.g., Morchella esculenta*). Another group of Ascomycetes includes the cup fungi, with a cuplike or dishlike fruiting structure, sometimes highly coloured.

Other unusual forms, not closely related to the true mushrooms, but often included with them, are the jelly fungi (*Tremella* species), the ear fungus or Jew's ear (*Auricularia auriculara-judae*), and the edible truffle.

World scenario of mushroom cultivation

Mushroom cultivation is a worldwide practice. World production of mushrooms over the last two decades has shown a phenomenal pattern of growth; with a 5 times increase in tonnage. The white button mushroom (*Agaricus bisporus*) still retains the highest overall world production, however, its relative contribution is decreasing due to the dramatic increase in the other species, *viz*: *Lentinus* and *Pleurotus* in particular. In terms of percentage, output yield of the leading 6 species, *viz*: *Agaricus bisporus* (31.8%); *Lentinus edodes* (25.4%), *Pleurotus* spp. (14.2%), *Auricularia auricula* (7.9%), *Flammulina velutipes* (4.6%), and *Volvariella volvaceae* (7.9%), cultivated made up to 87% of total world production. Overall, world production of mushrooms is increasingly being dominated by species that are both edible and have medicinal properties. The leading producers of white button mushroom (*A. bisporus*) are

France, China, Netherlands, United Kingdom and Taiwan; whereas the leading consumers of mushroom are Germany, Netherlands, Canada, France, United Kingdom, Sweden, USA, and Italy.

It is pertinent to note that world production of mushrooms is now dominated by China and it has become a major producer and consumer of both edible and medicinal mushrooms. Furthermore, China is also the major producer of the non-edible medicinal mushrooms, e.g. *Wolfiporia (Poria) cocos* and *Ganoderma lucidum*. At least 10 new species of edible or medicinal mushrooms have been brought into cultivation in China in recent years and although as yet on a small scale, the potential, especially for mushrooms of medicinal value, is quite significant. Because of their historical background in the use of wild mushrooms, both as food and in Chinese traditional medicines, it is to be expected that China would continue to develop methods for cultivation of an increasing number of, as yet, uncultivable mushrooms for medicinal exploitation. The traditional acceptance of mushrooms in herbal medicine and in expanding pharmaceutical industries will ensure that China will continue to be a major exploiter of medicinal mushroom technology.

Indian scenario on mushroom cultivation

Presently several varieties are cultivated in India out of which *Agaricus bisporus* comprised of 56% area followed by *Lentinus edodes* (14%), *Volvariella volvacea* (8%), *Pleurotus* spp. (7.7%) and others 13%. Out of total production of mushrooms in India, about 85% is met by the *Agaricus bisporus* (white mushroom) followed by *Pleurotus sajor-caju* (oyster mushroom) and *Volvariella volvacea* (paddy-straw mushroom). SSI-689 and SSI-17 are single spore selections of *Agaricus bisporus* mostly used for the cultivation of this mushroom.

Schedule of round the year cultivation of mushroom in northern India

1. February to Mid-April Oyster Mushroom (*Pleurotus sajor-caju*)
2. Mid June to Mid-September Paddy Straw Mushroom (*Volvariella volvacea*)
3. September to November Oyster Mushroom (*Pleurotus sajor-caju*)
4. Mid-November to Mid-March Button Mushroom (*Agaricus bisporus*)

Nutrition in Mushrooms

Mushrooms have significant nutritive value. Their chief worth is as a specialty food of delicate, subtle flavour and agreeable texture. Mushrooms are rich in proteins; contains less fat, less carbohydrate and salts; rich in fibre and have high Vitamin B_{12} and folic acid content so uncommon in vegetables; high availability of lysine and tryptophan amino acids usually absent in cereals; and, therefore, they are ideal food for patients suffering from hypertension, diabetes and obesity. By fresh weight, a common and commercial mushroom is more than 90% water, less than 3% protein, less than 5% carbohydrate, less than 1% fat (0.2 g/100 g), and about 1% mineral salts and vitamins.

Cultivated mushrooms consists of a good sources of vitamin B_2, niacin, and folates, with contents varying in the ranges 1.8—5.1, 31-65, and 0.30-0.64 mg/100 g dry weight, respectively. Compared with vegetables, mushrooms are also a good source of many mineral elements, e.g., the contents of K, P, Zn, and Cu varied in the ranges 26.7-47.3 g/kg, 8.7-13.9 g/kg, 47-92 mg/kg, and 5.2-35 mg/kg dry weight, respectively. *A. bisporus* is reported to contain a large amounts of Selenium (3.2 mg/kg dry weight), whereas, the levels of Cadmium is reported to be high in *Lentinus edodes* (1.2 mg/kg dry weight).

Mushrooms and health

Some forms of fungi and mushrooms are known to be beneficial for mankind as a nutraceutical. Recent studies have indicated that fungi could also be used in the development of medicines, medicinal foods, as well as food supplements.

Mushrooms are valuable health food as they are low in calories, high in vegetable proteins, chitin, iron, zinc, fibre, vitamins and minerals, Vitamin-B, phosphorus, and potassium. A handful of regular button mushrooms (*Agaricus*) can supply the entire vitamin B_{12} needed for the day. Modern researchers are focusing on the medicinal and healing capabilities of mushrooms.

The legendary effects of mushrooms on promoting good health, vitality and increasing our body's adaptive abilities as well, have been supported by numerous recent studies. Mushrooms assist our body to strengthen itself and it's immunity by maintaining physiological Mycoinfoostasis, and maintaining natural resistance to diseases.

The compounds that mushrooms contain have been classified as "Host Defence Potentiators" (HDP), which boosts our immune system. These disease-inhibiting compounds include hemicellulose, polysaccharides, nucleosides, triterpeniods, complex starches, and other metabolites. In combination they strengthen the immune system, aid in neuron transmissions, metabolism, hormonal balance, and the transport of nutrients and oxygen. Much work needs to be done to adequately characterise these molecules in the years to come.

Shiitake (*Lentinus edodes*) mushrooms are said to have quality of lowering blood pressure in those with hypertension, lower serum cholesterol levels, increase libido, and stimulate the production of interferon, which has vital anti-viral effects.

The Reishi mushroom (*Ganoderma lucidum*) can increase the production of interleukin 1 & 2, resulting in inhibition of tumour growth. It is also reported to have analgesic, anti-inflammatory, antioxidant, antiviral, antibacterial, expectorant properties and involve in lowering blood pressure, lowering serum cholesterol.

It is worth to emphasise that mushrooms support our systems best when we maintain a well-regulated body. Vitamins and mineral supplements should be taken on a regular basis with a well-balanced variety of fresh foods, good sleep, and positive daily attitudes and practices. Sometimes because of deficiencies or imbalances in these areas, it allows pathogens to attack our bodies, and we get sick. We already live in artificial environments where air is filtered and food is processed.

In nutshell, if we wish to keep our immune system functioning efficiently, we must add to out diet a natural immunity building supplements such as mushrooms.

Specialty mushrooms

The most popular mushroom consumed today is the white button mushroom. Because it dominates the market, many call anything other than the button mushroom a specialty mushroom. Following that logic, the Portobello is the most popular specialty mushroom, even though it is simply a brown variety of the button mushroom that is allowed to mature. Other popular specialty mushrooms are the Shiitake, Oyster, and Enoki.

Mushroom species for hot climate

There are several commonly cultivated strains of mushrooms that grow well in hot weather. The Paddy Straw mushroom (*Volvariella volvacea*) and its close relative (*Volvariella bombecina*), grow best at temperatures between 24-35 degrees C. The medicinal Reishi mushroom (*Ganoderma lucidum*), prefers warm weather (24 degrees C), and a strain of *Pleurotus ostreatus*, fruits at temperatures above 24 degrees C). The almond mushroom (*Agaricus subrufescens*), is a warm grower. The King Stropharia (*Stropharia rugosa-annulata*) also fruits only when temperatures rise.

Important mushrooms species and their temperature requirement for cultivation

S.No.	Species	Temperature (^{0} C) for mycelial growth		Temperature (^{0} C) for development fruiting body	
		Cardinal	Optimum	Cardinal	Optimum
1.	*Agaricus bisprorus*	3-32	22-25	9-22	15-17
2.	*A. bitorquis*	3-35	28-30	18-25	22-24
3.	*Auricularia auricula*	15-34	28	22-25	15-28
4.	*A. polytricha*	10-36	20-34	15-28	27-27
5.	*Flammulina velutips*	3-24	18-25	6-18	8-12
6.	*Hericium erinaceus*	12-33	21-25	12-24	15-22
7.	*Lentinus edodes*	5-35	24	6-25	15-20
8.	*Pleurotus numeko*	5-32	24-26	8-20	7-10
9.	*P. orstreatus*	7-37	26-28	12-30	25-30 tropical, 16-20 sub-tropical, 12-15 temperate
10.	*P. sajor-caju*	14-32	25-27	10-26	19-21
11.	*Tremella fusiformis*	5-38	25	20-28	20-24
12.	*Volvariella volvacea*	15-45	32-35	22-38	

Span of mushroom growing

The span of mushroom growing depends on several things, including the stage of mushroom growing one start with, the method of inoculation, the temperature, the kind of substrate being used, the mushroom species, and the specific mushroom strain. Starting at the very beginning, mushroom spores can take from a few hours to several days to germinate. A culture of mushroom mycelium growing on a Petridish of nutrient agar can take 24 hours (for morels) to upwards of a month (e.g. *Agaricus* spp. and *Stropharia Rugosa-annulata*) to spread across the better part of the plate. Using a chunk of agar culture to inoculate small jar of spawn, it can take 2-4 weeks for the spawn to reach maturity.

If one start with a quantity of spawn and fresh bulk substrate, it takes about two to three weeks for standard oyster mushrooms to reach fruiting stage. In short, the more optimum the substrate, the temperature, and any other relevant growing conditions, the faster the mushroom mycelium and the forming mushrooms will grow.

Mushrooms Phobia

The Mycophobia or mushroom phobia is a fear and aversion to Mushrooms. It is persistent, abnormal, and unwarranted fear of mushrooms, despite the understanding by the phobic individual and reassurance by others that there is no danger. The mushroom phobia is most common problem having significant impacts the quality of life. It can cause panic attacks and keep people apart from loved ones and business associates. Living with fear means one can never concentrate fully and give attention to work.

Mushrooms phobia, like other phobia, is created by the unconscious mind as a protective mechanism. Some past events related to mushroom emotions triggers mushroom phobia to develop. It may be linked to some points being triggered by myriad, benign events like movies, TV, or perhaps seeing someone else experience trauma.

Symptoms of mushroom phobia typically include shortness of breath, rapid breathing, irregular heartbeat, sweating, nausea, and overall feelings of dread, etc. The symptoms of mycophobia is altogether different in individual to individual everyone experiencing in their own way.

Since, mushroom phobia is a root disorder of phobia of any which cannot be cured by any kind of drug, in general. However, drugs given for the control, may suppress mushroom phobia temporarily through chemical interaction. Mushroom phobia can better be controlled by concentrating on work or some other things. Psychologists train the people of phobic in nature in different ways to put their attention in work.

Chapter 2

HISTORY OF MUSHROOM SCIENCE

During the era before 10, 000 BC, when hunting and gath ering were a part of every day life, women of the Americas did the gathering. Because they were supposedly blessed with the special ability to see better in dim light, they were successful in foraging for mushrooms and fungi along with young nettles, ferns, birch and willow shoots, and water weeds. The foods that women gathered were not just supplementary to the diet, but when hunting expeditions were unsuccessful, these foods were the staples. One could say that the women were successful in bringing home the mushrooms. During this era before fire was discovered, all foods including mushrooms were eaten in their simplest form, completely raw.

Seneca, a first century Roman philosopher, dramatist, and statesman, said of mushrooms, "*They are not really food, but are relished to bully the sated stomach into further eating*." Diderot, an 18th century French encyclopedist and philosopher was quoted in the Encyclopedie, "*Whatever dressing one gives to them, to whatever sauce our Apiciuses put them, they are not really good but to be sent back to the dung heap where they are born*."

Pliny, the Elder, a naturalist and writer of the first century, was acquainted with truffles, as were the Babylonians. Desert truffles were well known and highly revered in medieval Baghdad. These truffles might have come from the Kalahari Desert in Botswana, an area that is even known today for its wealth of truffle mines.

Pierre Francois de la Varenne, a chef of the French court during the mid 17th century, brought some modern touches to traditional French cooking of that era. He created a dish that remains classic in today's French cooking. Named for Marquis d'Uxelles for whom he worked is the French dish, *Mushroom Duxelles,* a preparation of minced mushrooms cooked in butter and seasonings, which is combined with vegetables, rice, or breadcrumbs to make a flavorful stuffing.

Mushrooms were gathered and eaten from the time of early man, yet it wasn't until the 18th century that the details of their cultivation were fully understood. Because mushrooms in the garden often appear to pop up overnight, even the ancient scholar Dioscorides thought them capable of spontaneous generation, a belief that held fast throughout medieval times.

There were early writers who could distinguish between poisonous and edible varieties but didn't include information on mushroom recipes until Paulet's book was published in 1793. More highly praised is a book written in 1841 by Roques whose work is still thought to be the best treatise on mushroom cookery to this day.

Charles Darwin relates his observation of a Patagonian mushroom. He tells us that the women and children of Tierra del Fuego, an island off the coast of Argentina, gathered fungus in great quantities and ate them uncooked. Other than a variety of berries growing on the island, the natives did not eat any vegetable other than the fungus.

Historically, mushrooms were gathered from the wild for consumption and for medicinal use. China has been the source of much early cultivation of mushrooms, e.g. *Auricularia auricula* (600 AD), *Flammulina velutipes* (800 AD), *Lentinus edodes* (1000AD) and *Tremella fuciformis* (1800 AD). *Agaricus bisporus* was first cultivated in France in 1600 while *Pleurotus ostreatus* was first grown in United States (US) in 1900.

First record on cultivation of Mushroom was found during the reign of Loius XIV (1638-1715). The earliest description for method of growing Mushroom was written by de Turnforte (a Frenchmen), which was published from Paris in 1707 by Royal Society of Science.

France was the leader in the formal cultivation of mushrooms. Some accounts say that Louis XIV was the first mushroom grower. Around this time mushrooms were grown during 1800 in special caves near Paris on horse manure, which set aside for this unique form of agriculture. The horse manure, which was stacked in, heaps and allowed to heat up naturally. This compost was used to grow spawn from meadow or mill tracts where horses had been trampling. From this spawn with a matter of chance, it was found growing *Agaricus bisporus*; which laid the method of growing white button mushroom.

From France, the gardeners of England found mushrooms a very easy crop to grow which required little labour, investment and space. Mushroom cultivation began gaining popularity in England with more experimentation. Moreover, English took advantage of natural supply of wild spawn in the country and exported to America, Germany, Denmark and Australia.

In the late 19th century, mushroom production made its way across the Atlantic to the United States where curious home gardeners in the East tried their luck at growing this new and unknown crop. However growers had to depend on spawn imported from England, which, by the time it reached the U.S. was of poor quality.

Mushroom cultivation reached the US in the late1800's with the use of imported spawn from England. Many problems arose from this imported spawn but they could reach the success in 1903 and the first producer of pure culture virgin spawn was the American Spawn Company of St. Paul Minnesota, headed by Louis F. Lambert, a French mycologist. He began the production of brick spawn and advertised it across the country as "Lambert's Pure Culture Spawn." It soon proved to be a very popular product. This spawn received a silver medal at the Universal Exposition in St. Louis in 1904. A measure of Lambert's success was that English spawn was soon being sold (1918) under the name "English Pure Culture Spawn." Since then, mushroom cultivation has well mushroomed.

In 1891, the first book on mushroom growing was published and it shed new light on the theory of cultivation. William Falconer, a mushroom grower and experimenter from Dosoris, Long Island agreed with the recommendations of agricultural journalists and compiled their theories in *Mushrooms: How to Grow Them; A Practical Treatise on Mushroom Culture for Profit and Pleasure.*

Mushroom cultivation now spans many centuries, it is only over the last 2-3 decades that there have been major expansions in basic research and practical knowledge leading to the creation of a major worldwide industry.

By the end of 19th century French mycologist Matrochot and Constantin (1894) succeeded in solving two important problems in mushroom cultivation, *viz.* discovered the cause of disease and its management through sulphur fumigation; and secondly they could germinate the spores in order to achieve sterile spawn. This was, probably, the most important contribution to scientifically controlled mushroom growing. In 1902, Miss Ferguson of Crone explained the technique of germinating the spores in detail. In 1905, American scientist Duggar invented the pure suspension culture from the mushroom tissues.

Surface cultivation of mushroom originated in Sweden, where Lundberg described mushroom growing in greenhouses in 1754.

In the later years French discovered the use of gypsum as it removes greasiness in the compost. Moreover, it was suggested by the Pizer, at Wye College in 1936 that gypsum immensely improves the spawn growth in compost, which is now universally accepted.

Lambert and Sinden were working on synthetic compost. In 1950, Sinden and Hauser introduced the short-term technique of composting. Lambart in 1929 introduced the technique of spawn production, whereas Sinden patented his grain spawn process in 1932. Flack in 1917 for the first time started growing *Pleurotus* species on stumps and logs of wood.

Folklore regarding Mushroom Science

Mushrooms were said to be the plant of immortality. That's what ancient Egyptians believed according to the Hieroglyphics of 4600 years ago. The delicious flavor of mushrooms intrigued the pharaohs of Egypt so much that they decreed that mushrooms were food for royalty and that no commoner could ever touch them. This assured themselves the entire supply of mushrooms. In various other civilizations throughout the world, including Russia, China, Greece, Mexico and Latin America, mushroom rituals were practiced. Many believed that mushrooms had properties that could produce super- human strength, help in finding lost objects and lead the soul to the realm of the gods.

Many cultures believed that eating mushrooms could endow them with super-human strength. This belief was prevalent in Russia, China, Greece, Mexico, and Latin America. Other beliefs concluded that partaking of mushrooms could lead the soul to reside with the gods.

Mushrooms grow upward with such surprising strength they are said to push up cement stones.

Some cultures believed that eating mushrooms gave them clairvoyance in locating lost objects.

Mushrooms were known to be such a powerful aphrodisiac that they became infused into important rituals and ceremonies. The Normans, during the 11th century, traditionally prepared a wedding dish that contained a pound of mushrooms to be fed to the groom only.

An Ancient Greek artifact from the Etruscan period, 8th century B.C., depicts unusual mushrooms, either the poisonous ***Amanita muscaria*** or ***Panaeolus papilionacceus,*** at the feet of Nessus the Centaur. Centaurs earned their reputation of possessing both divine wisdom and rather naughty behavior. Some believe their mischievous antics were enhanced by these potent fungi rather than by the ale and wine they consumed.

Of those who have studied Greek mythology, some believe that the ambrosia consumed during the Mysteries of Orphic and Eleusinian rituals was actually made from mushrooms, possibly the dung mushroom (***Panaeolus papilionacceus.***) Some scholars believe that these same mushrooms powered the fierce Norsemen known as ***berserkers*** who worked themselves into frenzy before battle.

In France truffles were unknown until the 14th century. As with many foods at that time that were unfamiliar to Europeans, truffles were attributed with powerful aphrodisiac abilities. The colorful legend of the Duke of Clarence tells that he married an Italian woman with a dowry of the hills of Alba, Italy, an area that was rich with truffles. On the night of his wedding he was said to have eaten so many truffles, referred to as "white diamonds," that he died before enjoying their magical abilities.

History of Mushroom Science in India

In 1886 some fine specimens of Mushroom were grown by N.W. Newton and exhibited at Annual show of Horticultural Society of India (Calcutta). Nutritional value of Mushrooms was explained by B.C. Roy in 1896 by collecting mushrooms at Calcutta Medical College, Calcutta. In 1908, a thorough research was instituted by Sir David Prain regarding edible Mushroom of India. Bose 1921 was first successful man to culture at least two Agarics Mushrooms on sterilized dung medium; the details of this finding were published in Indian Science Congress held at Nagpur (Maharashtra) during 1926. Bose and Bose 1940 discussed the method of growing mushroom on horse manure.

Experiments in cultivating the paddy straw Mushroom (*Volvariella volvacea*) were first undertaken in India by department of Agriculture Madras (Chennai) during 1939 and 1945. Su and Seth in 1940 outlined the procedure of spawn (*Volvariella volvacea*) production. Thomas & Co Workers in 1943 gave the detail method of cultivation of *Volvariella diplacea* in Chennai. Addition of *besan* (gram flour) was suggested by Asthana in 1947, for better yield of Paddy straw mushroom. Bano and Srivastava in 1962 described the cultivation of *Pleurotus* mushroom on straw based substratum including paddy straw.

Department of Horticulture (Himachal Pradesh) started a project on mushroom develop during 1977 under the guidance of United Nations Development Project (UNDP). The technical expertise was provided by project specialist James Tunney.

Mushroom cultivation in India was initiated for the first time at Solan in mid sixties when Dr. E. F. K. Mental from Germany started the work as the FAO consultant at Solan. He started the work on a small scale at the Dept. of Agriculture, H. P. Govt., Solan and successfully grew button mushrooms for the first time in India. Also associated with the project at Solan were the late Dr. P.K. Seth from the Dept. of Agriculture and Dr. S. Kumar. Simultaneously button mushroom cultivation was begun as a pilot project at Srinagar by Mr Stewart (a Britisher settled at Srinagar) along with a team of workers from the Dept. of Agriculture at Lalmandi, Srinagar. At both places the activity grew in size and mushroom growers started growing mushrooms in their houses as a cottage industry. Solan developed more rapidly as the effort at the Dept. of Agriculture (later Himachal Pradesh Agricultural University) was pursued vigorously and the first commercial unit of white button mushrooms in the seventies was put up at Kasauli (near Solan) by Mr. Saigal with the help from the C.O.A., Solan. Then came, his Highness, the Maharaja of Patiala who started growing mushrooms in the seventies at Dochi and Chail, near Solan on a much larger scale. He monopolised the button mushroom market in India for sometime and mushrooms from his farm would travel, in fresh condition, as far away as Bombay.

Meanwhile mushroom growing took the shape of a cottage industry in Kashmir in the seventies where people started growing button mushrooms on a large scale in villages on composts prepared by the long method. The mushrooms produced in

Kashmir were purchased by local canners and marketed in other parts of the country. The growers were producing button mushrooms against greater odds with little know-how available, especially for producing quality mushrooms. The mushroom activity spread to other hilly regions of India like the hills of Uttar Pradesh and Tamil Nadu.

Growers in the North Western (NW) plains started button mushroom growing (as a winter crop) to take advantage of the winter season in this region. Farmers in the NW plains in and around Delhi started growing a winter crop of button mushrooms successfully and marketing the produce in nearby Delhi.

In the mid-seventies Dr. W. A. Hayes was appointed as FAO expert on mushrooms in India and worked at Solan for some time to establish and standardize the facility for the short method of composting at the College of Agriculture, Solan. This resulted in the establishment of an air-conditioned cropping house at Solan with a facility for peak heating of mushroom compost.

With the contribution of the College of Agriculture at Solan, many students pursued a mushroom programme for obtaining their MS and Ph.D. degrees under the guidance of Prof. R. L. Munjar, Prof. and Head of Dept. of Plant Pathology and Head of Mushroom Research Laboratory, Solan. This was shortly followed by the establishment of a mother composting unit at Solan in the early eighties under UNDP assistance where James Tunney from the UK worked for several years commissioning the mother composting facility at Solan. The seeded compost at the mother unit was provided to growers in and around the Solan hills to initiate mushroom cultivation. The mushrooms thus produced were purchased by canners and sold in the market. But the consumption of mushrooms was confined to the richer sections of society.

In late eighties and early nineties some modern air conditioned mushroom farms were built near cities like Pune, Bombay, Delhi, Chandigarh, Hyderabad and Madras. With the introduction of spawn of improved/hybrid strains in the nineties by some leading spawn companies from the US and Europe, the industry got a further boost. By the early nineties production of all types of cultivated mushrooms was 5-7 thousand tonnes, and a minor part of it was exported. Besides that, mushrooms harvested from nature like Morels/Kabul Dhingri were exported in greater quantities to Europe and America in the dried form.

A serious attempt was made by department of agriculture, Himanchal Pradesh in collaborations with Indian Council of agriculture Research (ICAR) in 1961. In result this attempt, a scheme named "development of Mushroom cultivation in Himanchal Pradesh" was started. During 1964, Council of Scientific and Industrial Research (CSIR), New Delhi and Jammu and Kashmir government started production of white button mushroom in the valley. In 1965, Food and Agriculture Organisation (FAO) through its specialist E.F.K. Mental and W.A. Hays gave detailed technical help for development of mushrooms in India.

Commercial production of Mushroom was started from New Delhi and Solan (Himanchal Pradesh), and later spread to Jammu and Kashmir, Nilgiri, Punjab, Harayana, Chandigarh, Uttar Pradesh, Maharastra, Madhya Pradesh and Gujarat.

The ICAR created a center named "National Center for Mushroom Research and Training" (NCMRT) for the development of Mushroom industry in India during VI plan in October 1982, headquarter at Solan (Himanchal Pradesh). The objective of this centre was to conduct research on mushroom production problems, preservation, utilization and also to impart training to scientists, teachers, extension workers and progressive growers of mushroom. ICAR has also sanctioned "All India Co-ordinated Mushroom Improvement Project" (AICMIP) at six different centers for quick evaluation of mushroom technology. Centers started functioning from June 1983.

In this span of time NRCMT, Solan has developed single spore isolates (SSI) and hybrids of *Agaricus bisporus* out of which two SSIs NCS-100 and NCS-101 and a hybrid NCH-102 have been released for the benefit of Indian growers. Moreover, two superior strains of summer white button mushroom (*Agaricus bitorquis*) *viz.* NCB-13 and NCB-6, which are suitable for cultivation in plains and low hills, have been identified.

In India, cultivation technology of shiitake mushroom has also been standardized and will be available for growers in the near future. It has been established that using calcium chloride spray may enhance storage quality of the mushrooms.

The Centre has identified a fungicide sporgon which is very effective in controlling the diseases like *T. viridi* , *V. fungicola* , *C. dendroides* and *S. chrysospermum*. Four new mites, of white button mushroom namely *Histostoma* spp., *Tyrophagus* spp., *Neoselius* spp. and *Lcidonychus* spp., have been identified which would help in bio-control of insects.

The Centre has succeeded in increasing the yield of white button mushroom by spraying the hormone veradex. Also, Centre has developed technologies for production of many value added products of mushrooms *viz.* dried, pickle, soup powder, etc.

As ICAR has made it mandatory to have course curriculum on Mushroom cultivation in all the Universities engaged in imparting education on Agriculture. Therefore, all the agricultural universities are engaged in mushroom research and/or teaching. In addition, Indian Agricultural Research Institute (IARI), New Delhi and Indian Institute of Horticultural Research, Bangalore, is also engaged on research of quality mushroom.

Chapter 3

NUTRITIONAL VALUE OF MUSHROOM

One cup of sliced raw button mushroom contains, Calories (18), Total fat (0.3 g), Polyunsaturated fat (0.1 g), Dietary fiber (0.8 g), Protein (2 g), Carbohydrate (3 g), Sodium (3 mg), Riboflavin (0.3 mg), Niacin (2.9 mg), Pantothenic acid (1.5 mg), Copper (0.3 mg), Selenium (8.6 mcg). It is worthwhile to mention that mushroom doesn't contain Mono and Polyunsaturated fats, as well as cholesterol; hence they are useful for diabetic and heart patients.

There is saying, *"Whether you are counting carbs or calories or just trying to include more healthful foods in your diet, fresh mushrooms are ideally fit"*.

Mushrooms add satisfying flavour to food: Enjoying food is crucial for sticking to any weight loss plan and mushrooms provide a dual flavour punch. Not only every variety of mushrooms has its own distinctive taste, but also all mushrooms contain a unique flavour-boosting property.

Mushrooms make a meal extra-special: The sumptuous taste of fresh mushrooms gives most any dieter the feeling of indulging in a luxurious treat.

Mushrooms break mealtime monotony: With a selection that ranges from delicate white mushrooms to meaty portabellas to exotic shiitakes, fresh mushrooms remove boredom an obstacle to weight loss. And this ingredient is right at home with global flavours, whether one is interested in cooking a "comfort" meal or going global with Italian, Chinese, Thai or Tex-Mex.

According to Lintzel 1941, 100-200 g of Mushroom are required to maintain nutritional balance in a normal human being weighing 70 kg of body weight. He equated the nutritive value of mushrooms to that of muscle protein. Moreover, researchers have proven that mushrooms are well suited to supplement diets lacking protein, hence they have called mushroom as "*vegetable* meat".

Comparison of mushrooms with some common vegetables per 100 g of article (Wooster, 1954)

Name	Calories	Moisture (%)	Fat (%)	Carbohydrate (%)	Protein (dry wt. basis)
Mushroom	**16**	**91.1**	**0.3**	**4.4**	**26.9**
Beet root	42	87.6	0.1	9.6	12.9
Brinjal	24	92.7	0.2	5.5	15.1
Cabbage	24	92.4	0.2	5.3	18.4
Cauliflower	25	91.7	0.2	4.9	28.8
Celery	18	93.7	0.2	3.7	20.6
Green beans	35	88.9	0.2	7.7	21.6
Green peas	98	74.3	0.4	17.7	26.1
Lima beans	128	66.5	0.8	23.5	22.2
Potato	83	73.8	0.1	19.1	7.6

Energy Value of Mushroom: Mushrooms are good source of energy as 454 g of fresh mushrooms provide 120 K cal of energy (Hayes and Haddad, 1975).

Protein

Depending upon the variety, mushrooms contain 1 to 3% of protein and all the essential amino acids, making the protein complete. For vegetarians, mushrooms make an ideal meat substitute.

Mushroom contains protein, which consist of various amino acids. According to Hayes and Haddad (1976) almost all the essential proteins necessary for the human body are present in the mushroom. Tryptophan and lysine are present in high concentration as compared to cystein and methionine. Moreover, mushroom protein has digestibility as high as to 72-83 percent (Lintzel, 1941). Mushroom is said to contain high digestibility enzymes.

Comparative chart for nutritional value of different edible mushrooms (Zakia, 1976)

Mushrooms	Moisture	Ash	Protein	Fat	Carbohydrate
Agaricus bisporus	89.5	1.25	3.94	0.19	1.09
Lepiota spp.	91.0	1.09	3.30	0.18	0.86
Pleurotus spp.	90.0	0.97	2.78	0.65	1.08
Pleurotus ostreatus	92.5	—	2.15	—	—
Termitomyces spp.	91.3	0.81	4.1	0.22	1.13
Volvariella diplasia	90.4	1.10	3.90	0.25	1.57
Volvariella volvacea	88.4	1.46	4.98	0.74	1.38

Carbohydrates

By fresh weight mushroom contains 0.95 percent mannitol, 0.28 percent reducing sugars, 0.59 percent glycogen and 0.91 percent hemicellose (Mc Connel and Esselsen, 1947).

Vitamins

Mushroom is reported to contain an excellent amount of vitamin B-complex (0.14 mg/kg), riboflavin and nicotinic acid; also it is an excellent source of panthogenic acid. Moreover, it is good source of thiamin, folic acid and ascorbic acid (Kezeli and Dzabaridze, 1944). However, some researchers including Anderson and Fellers (1942) reported that *Agaricus bisporus* doesn't contain vitamin A, D or E. Chanterelles, with their appealing yellow colouring, are the only mushrooms that contain beta-carotene and vitamin D.

There are reports that fresh mushroom contain more amount of vitamins than dried mushroom, as it is indicated that about 60 percent of vitamin is lost during drying.

Composition of Amino acids in 100 g of *Agaricus biporus* dried matter.

Alanine	2.40	Lysine	1.62
Arginine	1.90	Methionine	0.39
Asportic acid	3.14	Phenylalanine	1.55
Cystine	0.18	Protine	2.50
Glutamic acid	3.14	Serine	1.89
Glycine	1.20	Threoenine	1.48
Histidine	0.64	Tryptophan	3.94
Isoleucine	1.28	Tyrosine	0.78
Leucine	2.16	Valine	1.63

Riboflavin promotes healthy skin and good vision and is also involved in energy production. Mushrooms are a particularly rich source of riboflavin. A serving of crimini or portabella mushrooms supplies about one-quarter of our daily value of riboflavin.

Niacin is involved in helping our body to produce energy and regulating the digestive as well as nervous systems. Mushrooms are a best source of niacin.

Pantothenic Acid is involved in the production of hormones and also plays an important role in the nervous system. Mushrooms are a good source of pantothenic acid.

Vitamin contents (mg/100 g of dry weight) of important edible Mushrooms

Species	Niacin	Ascorbic acid	Thiamin	Riboflavin
Agaricus bisporus	55.7	81.9	1.1	5.0
Lentinus edodes	54.9	0.0	7.8	4.9
Pleurotus ostreatus	108.7	0.0	4.8	4.7
Volvariella volvacea	91.9	20.2	1.2	3.3

Fats

It is reported that cream coloured mushrooms contain more fat than those of white varieties. Generally, mushrooms are rich source of linoleic acid, an essential fatty acid.

Minerals

According to Anderson and Fellers (1942), *Agaricus bisporus* contains high amount of potassium, phosphorus, copper and iron. However, the calcium amount is comparatively low as against of the other minerals.

Fresh mushroom contains 0.9-1.7 percent of minerals which is greater than those of fruits and vegetables (except leafy vegetables). Button mushroom contains 0.45 percent phosphorus, 0.144 percent potassium, 0.11 percent sodium, 0.006 percent calcium, 1.55 percent copper, 18.7 ppm (part per million) iron, 13.2 ppm magnesium.

Potassium is vital in maintaining normal heart rhythm, fluid balance, muscle and nerve function. There is strong evidence that a diet that includes good sources of potassium can also help reduce the risk of high blood pressure and stroke. Mushrooms are an under-recognized source of potassium, for instance, a 3-ounce portabella mushroom cap provides more potassium than a banana or an orange.

Selenium is an essential trace mineral that supports normal growth and reproduction. Recent studies also suggest that selenium intake may help lower the risk of prostate cancer. Mushrooms are a good source of selenium. A serving of crimini mushroom supplies almost one-third of the daily value of selenium.

Copper is another essential mineral and mushrooms are a good source. We are all aware of iron's role in making red blood cells and delivering oxygen to every part of the body, but did you know that iron couldn't do its job without copper?

Chapter 4
CLASSIFICATION AND MORPHOLOGY OF MUSHROOMS

Nearly every amateur mycologist and many professional mycologists, especially those not engaged in systematics, see mushrooms much the way they have been understood, described and classified for over 200 years. The standard modern approach to the basidiomycetes has been that of Elias Fries' 1874 *Hymenomycetes Europaei*. Fries divided all the basidiomycetes into 6 major groups based on the shape of the fruiting body (sporocarp) in the field. These shapes include mushrooms containing (1) gills, (2) pores, (3) teeth, (4) vase, parchment or crust like fruiting bodies, (5) coral-like fungi, and (6) tremelloid or jelly-like mushrooms.

DNA sequencing and Cladistics in Mushroom Classification

In last 20 years new technologies and methodologies have been worked out in the mushroom science. The DNA sequencing techniques with PCR (polymerase chain reaction) including Cladistics (The systematic classification of groups of organisms on the basis of the order of their assumed divergence from ancestral species) have bee used for arranging and presenting data for mushrooms. Based on the above technologies, Phylogenetic trees (cladograms) attempting to show in a visual way the evolutionary relatedness among diverse fungi have been developed for many groups of mushrooms. Different regions of sequenced DNA or different methodologies for presenting this data, however, have led to different phylogenies (cladograms). David Hibbett, have used nuc-lsu rDNA (nuclear large subunit ribosomal DNA) sequences for constructing phylogenies of the homobasidiomycetes; Tom Bruns used the ITS (internal transcribed spacer region) molecule in his research for developing phylogenic tree. However, others still use conventional nuclear and mitochondrial small subunit and large subunit ribosomal DNA sequences for the same. Some sequences, *e.g.*, ITS, appear to be good at the species level for some groups of fungi; others, *e.g.*, nuc LSU, at higher taxa levels. Still others are used to discern trends in morphological evolution rather than for identifying or classifying mushrooms.

Silver bullet approach

In a sense, both the field-based and the DNA approaches rely on single-character taxonomy, the **silver bullet approach**, for mushroom identification and classification. Field-based taxonomy is based on the belief that we can identify mushrooms, much as can identify plants and animals, with no special education or equipment. DNA-based taxonomy is based on the belief that we can identify any organism by finding its bar code.

Lineage of Euagarics

Homobasidiomycetous fungi, including mushrooms and puffballs, display many complex fruiting body morphologies, but their anatomical simplicity has confounded efforts to understand the evolution of these forms. Scientists have performed a comprehensive phylogenetic analysis of homobasidiomycetes, using sequences from nuclear and mitochondrial ribosomal DNA, with an emphasis on understanding evolutionary relationships of gilled mushrooms and puffballs. Parsimony-based optimization of character states on our phylogenetic trees suggested that strikingly similar gilled mushrooms evolved at least six times, from morphologically diverse precursors. Approximately 87% of gilled mushrooms are in a single lineage, which we call the "**euagarics**." Recently discovered 90 million-year-old fossil mushrooms are probably euagarics, suggesting that (i) the origin of this clade must have occurred not later than the mid-Cretaceous age and (ii) the gilled mushroom morphology has been maintained in certain lineages for tens of millions of years.

Puffballs and other forms with enclosed spore-bearing structures (Gasteromycetes) evolved at least four times. Derivation of Gasteromycetes from forms with exposed spore-bearing structures (Hymenomycetes) is correlated with repeated loss of forcible spore discharge (ballistospory). Diverse fruiting body forms and spore dispersal mechanisms have evolved among Gasteromycetes. Nevertheless, it appears that Hymenomycetes have never been secondarily derived from Gasteromycetes, which suggests that the loss of ballistospory has constrained evolution in these lineages.

Some of edible mushrooms and their classification

Sub-division		**Basidiomycotina**	
Class		**Hymenomycetes**	
Sub class		**Phragmobasidiomycetidae**	
Order		Tremellales	
Family	1.	Tremellaceae	Genus *Tremella*
Family	1.	Tuberaceae	Genus *Tuber*
Order		Auriclareales	
Family	1.	Auricularaceae	Genus *Auricularia*
Sub-Class		**Holobasidiomycetidae**	
Order		Aphylophorales	
Family	1.	Pleurotaceae	Genus *Volvariella*
	2.	Hericeaceae	Genus *Hericium*
Order		Agaricales	
Family	1.	Agaricaceae	Genus *Agaricus*
	2.	Coprinaceae	Genus *Coprinus*
	3.	Polyporaceae	Genus *Lentinus*
	4.	Trichomataceae	Genus *Pleurotus*
	5.	Strophariaceae	Genus *Tricholoma, Flammulina, Calocyble, Stropharia, Pholiota, Hymenoloma*

Habitat

Generally edible mushrooms grow like a saprophyte in the field, pastures and open fallow lands close to the forest. The land with a thin forest crops is very suitable for this species but is never found in a dense forest as well as in coniferous forests. The mycelium is perennial, which consist of branch like spreading colourless hyphae. The cells of hypha contain granular vacuole and protoplasm with many nuclei and oil drops. These serve as food for the fungus. The hyphae of *A. bisporus* are not connected with each other, i.e. they remain separated but very close to each other. Generally in the month of September to October the temperature and humidity attain a level which conducive to forming sporophore or pinhead.

Based on the habit, mushrooms may be divided into following categories:

1. **Humicolous (humus inhabiting) mushrooms**

i). **Saprophytes,** e.g. *Lepista nuda, Volvariella* spp., *Marasmium* spp., *Polyporus* spp., etc.

ii). **Symbionts,** e.g. *Tricholoma* spp., *Lactyrius* spp., *Boletus* spp., etc.

2. **Wood inhabiting mushrooms**

i). **Saprophytes,** e.g. *Agrocybe* spp., *Pleurotus* spp., *Auricularia* spp., *Lentinus* spp., etc.

ii). **Symbionts,** e.g. *Armilaria mellea, Cyttaria* spp. etc.

3. **Dung inhabiting mushrooms, e.g.** *Agaricus* spp., *Coprinus* spp., *Agrocybe* spp., *Armellaria mellea, Pleurotus* spp., etc.

Type of sexuality in important mushrooms

Mushrooms species	Type of sexuality
Volvariella volvacea	Primary homothalism
Agaricus bisporus	Secondary homothalism
A. bitorquis	Heterothallism
Pleurotus ostreatus	Bifactorial homothalism
Pholiota nameko	Unifactorial heterothallism
Auricularia auricula	Unifactorial heterothallism
Lentinus edodes	Bifactorial heterothalism

MORPHOLOGY OF MUSHROOMS

Mushroom in common refers to the fruiting body (sporophore) of gill fungi. Mushrooms are characterized by the presence of spore bearing layer known as *Hymenium*. This layer may be present either side of gills as it is present in *Agaricus* spp. There are following different parts of a mushroom in order of top to bottom:

Fig. Growth and Development of Fruiting body

The cap or Pileus

It is an expanded portion of the sporophore which may be thick fleshy membranous or corky and varies in different shape and size and also in colour. The surface of pileus may be smooth, hairy or rough.

Gills or Lamellae

They are situated on the under side of the pileus starting from the apex of the stalk and radiating out towards the margin. The gills bear spores on their surface and exhibit a change in colour corresponding to that of spores. For e.g. in *Agaricus bisporus* the colour of the centre of the gills is pink but in due course of time colour changes to dark purple, brown nearly black. The change in colour of gills is due to the change of colour of spores moving towards maturity.

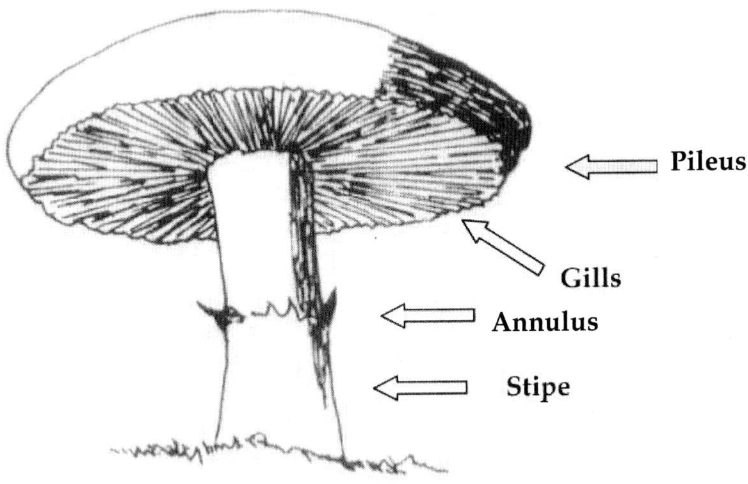

Fig. Sporophore of *Agaricus bisporus*

Based on attachment of gills to the stipe, they are classified in different names and help in identification of the mushroom species. For instance, when gills do not touch the stalk or if touches, it is with fine hairs, such situation is called **free** attachment. They are called **adnate**, if the attachment is directly to the stalk. The **decurrent** situation is so called when the gills extend down the stipe either greater or lesser is degrees. When are near the stalk in deep notch they are termed as **sinuate**.

Structure of a gill: The centre of a gill is made up of threads known as trama. Outsides of the trama cells branch into short cells forming a thin layer called *subhymenium*, which give rise to long club-shaped cells called basidia. Each basidium has 2-4 spine projections called sterigmata on which basidiospores are borne.

Basidiospores could be of different colours, shapes having either rough or smooth margins. The tiny spores are not visible with naked eye, appearing in dusty form. Approximately 16 billion spores are formed in each sporophore at one go. Upon suitable environmental conditions, spores germinate with germ tube giving rise to hypha (pl. hyphae), which after networking form mycelium.

Among the basidia there are several sterile cells called cystidia. Parallel to the basidia there are other sterile cells called paraphyses. The three, basidium, cystidium and paraphysis, constitute the hymenium.

Veil

The gills of young fruiting bodies remain covered by a delicate tissue extending from margin of cap to the stipe, is called **veil**. As the cap expands the tissue, being delicate, breaks away leaving the some remnants attached to the margin of cap and others to the stipe, forming a ring like structure known as **veil annulus**. This annulus is absent in *Volvariella* species.

Stipe or Stalk

The stipe is an organ supporting the cap. Its presence or absence makes the important character for the identification the particular spp. Usually stalk is centrally attached to the cap but in some cases it is also attachment in different ways. If attachment is either side but not in the central core, it is said to be **eccentric**.

Fig. Breakage of veil

Stipe may be hollow pithy, solid and fleshy in structure. The shape of stalk may be cylindrical having same diameter throughout the stalk or spindle (swollen in the middle and tapering towards the ends) shaped or club (enlarged at the top and tapering towards the base) shaped.

Universal Veil (Volva)

Prior to differentiation entire fruit body is covered by a tissue called universal veil. As the fruit body extends, the veil breaks away leaving a cup (the volva) surround the base of the stipe. It is not always, but some of the remnants in the form of scales remain hanging from the pileus. Presence of volva is the unique feature of *Volvariella* species. This usually absent in *Agaricus bisporus*.

Presence or absence of volva has special meaning in identification of mushroom species, as:

1. Both annulus and volva present, e.g. *Amanita* spp.
2. Only annulus present but volva absent, e.g. *Agaricus* spp.

3. Only volva present and annulus is absent, e.g. *Volvariella* spp.
4. Both volva and annulus are absent, e.g. *Marasmius oreades*

Rhizoids

Rhizoids are the organ of food absorbing usually attached below corresponding to the stalk/stipe.

Chapter 5

SUBSTRATES AND THEIR MANAGEMENT

The raw materials available for mushroom cultivation vary in different parts of India. In most of the northern and central parts of India, wheat straw is widely available but at high prices. In the eastern and southern parts, paddy straw is available in abundance, and at lower prices. Poultry manure is available everywhere at a very low price. Sugarcane bagasse is available in those areas where sugarcane is widely grown (western parts of India, central India, and some other places).

The requirements and techniques for composting differ with different base materials (wheat straw, paddy straw/ or s ugarcane bagasse) and information available to growers on the use of paddy straw/ or sugarcane bagasse as a base material is limited. Optimal utilization of the above materials for composting to obtain economic returns requires specialized skills from the manager/entrepreneur.

Casing is the second important input in button Mushroom growing, in substrate point of view. A very few casing materials are available for commercial growing in India, the choice has even narrowed down to Farm Yard Manure (FYM)/spent compost/ composted coir pith, etc. These materials require to be processed (water leached/steam pasteurised) before use, unlike the peat sold in Europe that is harvested from underground bogs deep down. This material is devoid of harmful microorganisms and is used after adjustment of pH with lime. Use of the above agrowastes as casing in button Mushroom cultivation, in place of peat, again requires experience on the part of the grower for their optimal utilisation.

Commercial mushrooms production can be achieved in specially formulated and processed compost made from a combination of various materials including horse and chicken manure.

Horse manure has long been used, but other materials are also suitable for mushroom cultivation. Some materials that have been used alone or in combination with horse manure are alfalfa hay, maize cobs, cornstalks, straw, sawdust, dried brewer's grains, etc.

Important ingredients and their uses

Ingredient	Purpose of use
Maize cobs	Nitrogen, Carbon and to add bulk
Hay	Nitrogen, Carbon and to add bulk
Horse manure	Nitrogen, Carbon and to add bulk
Straw	Carbon and to add bulk
Ammonium nitrate	Nitrogen
Animal fat	Nitrogen
Brewers' grain	Nitrogen
Dried blood	Nitrogen
Feathers	Nitrogen
Fish soluble	Nitrogen
Licorice roots	Nitrogen
Live stock manures	Nitrogen
Seed meal	Nitrogen
Seed oil	Nitrogen and carbon
Urea	Nitrogen
Ground wallboard	Flocculent and pH control
Gypsum	Flocculent and pH control
Lime	Flocculent and pH control
Spent lime	Flocculent and pH control

Hardwood bark	Carbon and to add bulk
Tree leaves	Carbon
Peat moss	For casing
Super phosphate	To add phosphorus
Muriate of Potash	To add potassium
Potato waste	Carbon
Seed hulls	Carbon
Shredded newspapers	Carbon
Bagasse (Sugarcane)	Carbon and to add bulk
Molasses	Carbon

Substrates for specific mushroom species

Growing Substrate	Mushroom Species
Banana leaves	*Volvariella* species (Straw mushroom)
Bean straw	*Pleurotus* species (Oyster mushroom)
Cocoa shell waste	*Pleurotus* species (Oyster mushroom)
Coffee pulp	*Pleurotus* species (Oyster mushroom) *Lentinus* species (Shiitake mushroom)
Coir	*Pleurotus* species (Oyster mushroom)
Maize cobs	*Pleurotus* species (Oyster mushroom) *Hericium* species (Lion's Head mushroom) *Lentinus* species (Shiitake mushroom)
Cotton seed hulls	*Pleurotus* species (Oyster mushroom) *Lentinus* species (Shiitake mushroom)
Cotton straw	*Pleurotus* species (Oyster mushroom)
Cotton waste from textile industry	*Pleurotus* species (Oyster mushroom) *Volvariella* species (Straw mushroom)
Crushed bagasse and molasses wastes from sugar industry	*Pleurotus* species (Oyster mushroom)

Distillers grain waste	*Hericium* species (Lion's Head mushroom)
Horse manure (fresh or composted)	*Agaricus* species (Common button mushroom)
Logs	*Pholiota* species (Nameko mushroom) *Lentinus* species (Shiitake mushroom) *Tremella* species (White Jelly mushroom)
Oil palm pericarp waste	*Volvariella* species (Straw mushroom)
Paper	*Pleurotus* species (Oyster mushroom) *Stropharia* species
Rice straw	*Volvariella* species (Straw mushroom) *Pleurotus* species (Oyster mushroom) *Agaricus* species (Common button mushroom)
Sawdust	*Lentinus* species (Shiitake mushroom) *Pleurotus* species (Oyster mushroom) *Hericium* species (Lion's Head mushroom) *Auricularis* species (Ear mushroom) *Ganoderma* species (Reishi mushroom) *Grifola frondosa* (Maitake mushroom) *Flammulina* species (Winter mushroom)
Sawdust-rice bran	*Pholiota* species (Nameko mushroom) *Auricularis* species (Ear mushroom) *Coprinus* species (Shaggy Mane mushroom) *Flammulina* species (Winter mushroom) *Lentinus* species (Shiitake mushroom)
Sawdust-straw	*Pleurotus* species (Oyster mushroom) *Stropharia* species
Water hyacinth/Water lily	*Pleurotus* species (Oyster mushroom) *Volvariella* species (Straw mushroom)
Wheat straw	*Pleurotus* species (Oyster mushroom) *Agaricus* species (Common button mushroom) *Stropharia* species *Volvariella* species (Straw mushroom)

Shiitake Mushroom Cultivation in Enriched Sawdust Culture

In this innovative approach various hardwood sawdust or wood chips supplemented with nitrogen-rich additives such as rice bran are mixed together and then compacted into special autoclavable polypropylene bags of various dimensions. The bags are then autoclaved at 15 PSI (pounds per square inch) for 2 to 4 hours to ensure complete internal sterility, allowed to cool to 20 ^{0}C and then aseptically inoculated with the desired amount of spawn. The inoculated bags, sometimes known as **space bags** or artificial logs, are placed to growing rooms with controlled environments giving accurate humidity and temperature conditions.

A common recipe for supplemented sawdust is: 76% sawdust, 12% millet, and 12% bran, and 65% moisture. Note that recipes are always based on dry weights of the ingredients because moisture contents of the ingredients vary.

Overall, this new technique greatly shortens the production time and gives much higher yields as compared to natural log cultivation. A 100 kg sawdust plus supplement can produce 80 kg fresh weight of mushrooms representing 8-10 fold greater production than the natural log method.

Various other formulae for sawdust culture

Sawdust from *Quercus crispula*	80%
Rice bran	20%
Water contents adjusted to	65%
	Mori *et al.*, 1974

Sawdust from hardwood	80%
White millet	10%
Wheat bran	10%
Water to	4 ltr
	Royse *et al.*, 1981

Sawdust from hardwood	89.8%
Rice bran	10%
Calcium carbonate	0.2%
Water to	60%

Ham *et al.*, 1981

Sawdust from hardwood	81.2%
Rice bran	12%
Corn/rice/millet powder	6%
Calcium carbonate	0.8%
Water to	62%

Song, 1986

Mushroom Cultivation in Cotton Waste

Cotton is still the worlds leading textile fibre. Waste, generated from the mechanical processing of raw cotton prior to spinning, provides an ideal substrate for the growth of some edible mushrooms *viz. Volvariella volvacea, Pleurotus* spp. About 7% of lint (i.e. fibre) waste is produced in spinning.

Volvariella volvacea cultivation in cotton waste

S T Chang from Chinese University, Hong Kong, is pioneered for the development of cotton waste as a substrate for the indoor cultivation of *V. volvacea.*

Cotton waste has the advantage of being much cheaper and readily available materil giving a higher and more stable yield of mushrooms, generally 30-40% Biological Effieciency (B.E.), and earlier appearance of pinheads (5 days after spawning) as well as crop (first harvest 3-5 days after pinheads).

The cotton waste is mixed with 4-5% rice or wheatbran and about 2-6% limestone (or calcium carbonate). It is important to partially compost the mixture and this process, following the addition of water, is carried out in a wooden frame for about six days with mixing at 2 day intervals. The compost is then made up into beds and pasteurized by introducing steam into the mushroom house. An air temperature of 60-62 ^{0}C is maintained for 2-3 hrs and the mushroom house is then cooled to 52 ^{0}C by the introduction of fresh air. This temperature is maintained for up to eight hours with a continued stream of fresh air. The steam valves are then closed and the air temperature allowed to drop gradually to 34-36 ^{0}C for spawning. This latter step takes about 12-16 hrs depending on the outdoor temperature. The temperature of the room is maintained at 32-34 ^{0}C during the period of spawn running.

Straw mushrooms are not usually left to grow to their maximum size but are picked at the stages before the volva enclosing the cap breaks or just after rupture. The former is referred to as the 'button stage' and the latter the 'egg stage'. The first flush lasts for about four to five days and is followed by a second flush about a week later. However, the yield of the second flush is only about 10% of that of the first one and most mushroom growers harvest only the first flush.

Straw mushrooms are highly perishable and must be marketed within 1-2 days. They tend to liquefy when refrigerated but have a shelf life of about 3 days when stored at room temperature.

Pleurotus spp. cultivation in cotton waste

Generally, commercial cultivation of oyster mushroom is carried out on straw, but now some non-composted materials such as cotton waste supplemented with rice bran and calcium carbonate is used as an effective substrate. Cotton waste has the potential to yeild mushroom (*Pleurotus pulmonarius*) up to 100% of its biological efficiencies. Cotton waste is supplemented with 5% wheatbran for better results.

Mushroom Production on Ligno-Cellulose

Mushroom cultivation is now possible in ligno-cellulosic wastes such as waste paper, cotton waste, straw, wheat, or rice bran. In the lignocellusic wastes, mushrooms like *Volvariella volvacea,* and *Pleurotus sajor-caju* have effectively been cultivated.

For compost preparation, paper substrates are shredded. Straw is either trimmed, coarse ground, or used directly. Five per cent of wheat or rice bran, and 5 percent calcium carbonate ($CaCO_3$) are added, along with sufficient water to raise the moisture content to about 60 per cent. This requires that approximately 1,500 ml of water be added per kg of ligno-cellulosic waste. The substrate should then be steamed for 30 minutes, before spawning.

Log Culture for shiitake mushroom production

The logs (3′ long x 6-8″ diameter) are cut in winter or early spring from fast growing deciduous species, which have thick outer bark. Logs are then inoculated with spores or with mycelial plugs inserted into drilled holes. The inoculated logs are stacked in piles with mild to heavy soaking. Mycelial growth through the log will occur over several months and the logs are then placed in an upright position partly embedded in the soil. Mushroom production will then occur primarily in the cooler spring and autumn months. Since this is an open, non-sterile procedure contamination with other wood rotting mushroom species can also occur. Increased environmental control can be achieved by having overhead protection or even growth in greenhouses.

Mushroom Cultivation with Hydrogen Peroxide

Hydrogen peroxide simplifies the whole process of growing fungi. There's no need to build a sterile laboratory, a low concentration of peroxide keeps out the contaminants, while allowing healthy growth of mushroom tissue. And as the mushroom tissue grows, it converts the peroxide to water and oxygen, leaving a clean, vigorous mushroom culture. In nutshell, with the hydrogen peroxide, mushroom cultivation can done in the open place without any contamination.

MANAGEMENT OF RAW MATERIALS AND MUSHROOM SUBSTRATE

Generating a substrate for the growth of the mushroom crop is a critical first step in mushroom culture for most cultivated commercial mushrooms. Poorly produced compost will invariably mean disappointing yields for the farmer. Because some materials used in this substrate are high in nutrients, the mushroom farmer must take care at the farm to keep these nutrients separated from the natural water resources. The management of substrate should be in such ways, which help protect these resources from degradation.

A well-run operation for preparing substrate should evidence an efficiency of design and orderliness of operation which, in concert with good housekeeping and water management, will go a long way to protect water and air resources.

Watering Raw Materials and Pile

Watering the raw materials and the piles of substrate is an essential operation in composting. If watering is optimal, bacteria and fungi cause changes in the raw materials in the pile, changing the mix of raw materials into a substrate that is selective for the Agaricus mushroom.

Bacteria and fungi also metabolize nutrients and convert them into microbial protein. Excess water, however, can lead to anaerobic activity within the heap and produce a sour product with the risk of malodours. Excess water can also lead to the loss of nutrients, which can become pollutants if not properly controlled.

A recommended practice is to add sufficient water, but not cause washout of ingredients and generation of large quantities of leachate. To find the balance point, wetting of the initial piles of substrate ingredients in the preliminary or "pre-wet" step, the watering volumes and schedules for the substrate heaps, heap turning schedules, and composting period can all be modified.

Since loss of some water from the piles is normal and unavoidable, the design and operations of the compost platform must be adequate to control the highly nutritious water. This objective, simply stated, is to keep all leachate, whether from raw ingredients or from the heaps, over the platform surface unless it is being discharged to an approved wastewater collection unit or recycling system.

The platform should be of sufficient size to accommodate the heaps and also the movement of raw materials. Storage locations should be chosen taking into account the concentrations of soluble nutrients in the raw materials and the nutrients' potential to generate leachate of environmental concern. Placement of raw materials and heaps should be optimized to reduce unnecessary traffic and tracking of materials on the wharf. It is critical to provide sufficient room for operations and high quality storage areas for raw materials.

The composting area must be covered with an impervious material and sloped toward impervious collection areas. Though roofing is expensive, it is advisable for storing high nutrient, leachable raw materials like chicken manure. Roofing will minimize the potential for contact of manure nutrients to contact with groundwater and surface water and thus minimize the leaching of nutrients into the collection basin.

While storage of raw materials into a covered heap is the preferred management practice, alternatives can be considered for certain raw materials that have minimal or no potential to leach nutrients or other constituents. For these materials, which do not generate leachates, bare ground storage is acceptable. For added protection, the raw materials should be covered securely with waterproof tarpaulin. In some cases, inventory quantities can be reduced to accommodate the size of the storage area.

Management of Waste

Although the farmer's goal may be to generate zero waste material, a small amount of waste may be unavoidable. Wastes generated in the mushroom substrate production process, for example, wastewater, excess substrate or off-spec substrate must be collected, stored and reused. If these wastes are not directly reused, they must be stored in accordance with the residual waste regulations.

Odour Reduction

Malodours are sometimes associated with compost operations. They are a natural occurrence, but need to be minimized. There are some ways the mushroom farmer can reduce the odour potential of the heap area. The following practices may help controlling and reducing odours:

1. Do not use more nitrogen-rich raw ingredients in the compost than necessary to produce nutritious compost for the future mushroom crop.
2. Maintain aerobic composting operations. To maintain aerobic condition, do not over water the heap.
3. Make sure the initial mix of ingredients is thorough, especially with manures, to avoid producing anaerobic clumps in the compost.
4. Prevent standing water and poor drainage areas in the composting platform, including around raw material storage areas.
5. Trap solids before they enter the collection basin. If solids are already present in the basin, remove them regularly.
6. Maintain maximum available isolation distances from where the composting operation is carried out and other potential land uses.
7. Screening (fences, trees, landscaping,) may be helpful in shielding the composting operation from the view of adjoining property owners.

Chapter 6

SPAWN AND SPAWNING

White mushrooms, like all mushrooms, grow from microscopic spores, not seeds. Plants growing from spores are called fungi. Mushroom spores are produced in the gills. As the mushroom matures the cap opens and the gills are exposed. Spores are microscopic spheres roughly comparable with the seeds of higher plants. These spores are produced in large numbers in the gills. A 7.5 cm mushroom produces as many as 40 million spores an hour. Spores are collected in the nearly sterile environment of a laboratory and then used to inoculate grains or seeds to produce a product called spawn.

The grower must have the clear idea about the seed of Mushroom. Since Mushroom is a fungus and does not bear flowers. Therefore, in actual sense seed is not produced. Instead of seed spores are produced which are used as a seed for propagation.

THE SPAWN

The term Spawn has been derived from the Latin word, ***expandere***, which means to spread out. It is a fungal propagule used for propagation of new mushroom crop. In other words, it is equivalent to seed of a plant (having no embryo) used for propagating mushroom.

The spawn is maintained in culture bank in a refrigerator (through frequent sub-culturing), in liquid paraffin and liquid nitrogen (cryopreservation). Spawn is prepared and supplied in 500g and 1kg polypropylene packs and one kg of mushroom spawn costs about Rs.50.00.

Availability of Spawn

The availability of spawn of various strains is limited to S-11, 310, 791, 76 (all non-hybrid strains of *A. bisporus*) and U-3 (Hybrid). In addition to these, NRCM has also released 3 new strains (NCS-100, NCS-101 and NCH-102) of *A. bisporus* in 1997 and 2 strains of *A. bitorquis* (NCB-6 and NCB-13) in 2000. The cultures of all these strains are available from the culture bank maintained at the centre at NRCM, Solan.

Presently Indian market demand is about 8000-10000 tones of spawn. Spawn in India is mostly produced on wheat grain and bajra grain (lesser millet). The spawning rate in button mushroom is 0.5-0.7% and in *Pleurotus* is about 2% of the wet weight of substrate.

Historical Background

Spawn was first obtained by digging up wild mycelium from meadows where wild agarics grew and horses were active. The early methods of finding, digging up perpetuating the wild mycelium were uncertain and unreliable. The first manufactured spawn combined a mixture of horse and cow manure pressed into bricks, the original source being the wild mycelium. The bricks were not sterilized, however, and could harbor pathogens and weed molds.

In 1915 pure culture spawn was produced on sterilized horse manure compost in bottles. The development of horse manure spawn, however, did not eliminate the problem of obtaining productive, reliable spawn, free from pests.

In 1930, The Pennsylvania State College (University) employed Dr. James W. Sinden to work on mushroom production problems. During his first series of experiments Dr. Sinden was struck with the variation in yields and feared that this might be a result of spawn planted. He realized that in order to carry on further experimental work on mushroom growing, it was essential to eliminate any variation from this cause. Therefore, Dr. Sinden sought a medium on which the mycelium would grow more vigorously and that would provide a more uniform product. One of the first of these used was grain, specifically wheat, which was placed in flasks with a small amount of water and heat sterilized. On introduction of the mycelium he found that it grew very vigorously and in a manner entirely different than anything previously seen. Thus grain spawn was born and is used almost exclusively today.

Various types of spawns

1. **Virgin Spawn**: This spawn is produced by sowing spores on a particular substrate, of required species. After growth, mycelium becomes thick with the inter-woven hyphae. Hypahe of this stage are used as spawn for spawning the substrate, is called spawn. This type of spawn was used for the first in France.

2. **Mycelium spawn:** Mushroom seeds are actually the sterile mycelium grown in a sterilized cereal starch medium containing bottles or polythene bags. It is prepared in the laboratory and the bottles of spawn are sold to the farmers. If the supply is not adequate than a substitute can be used. The Mycelium of the bed just after the harvest of the first batch is usually used for the purpose of spawning. The lumps of compost having more mycelium should often be collected. They should be cut in the pieces of small cubes and used as spawn immediately after collection. However, the Mycelium spawn is not properly pasteurized therefore it attracts several weeds.

3. **Flake Spawn:** When the compost of a bed is covered with a mycelium growth just before appearing the pinhead. The pile of compost is broken and mycelium is dried. When this dry mycelium with compost is used for spawning, it is called flake spawn.

4. **Brick Spawn**: A compost layer of 2-3 inch thickness is made with the mixture of horse and cow dung manure mixed with loam and water. Then the layer is cut into small brick like pieces. Into each brick a small depression is made and mycelium is put into them. The mycelium is kept for few days in the bricks under growth conditions. Now the fully-grown mycelium brick used for spawning, such spawn is called brick spawn.

5. **Grain Spawn:** In such type of spawn, grains are used as a substrate for growing mycelium. Particularly wheat or rye grains are used for the purpose. Grain spawn is a very poplar and used widely by the farmers of all level. While preparing the grain spawn, grains generally becomes clotted, which can be prevented by addition of slight amount of gypsum and chalk dust (6 g of gypsum and 1.5 g of chalk dust for 450 g of grains).

Spawn production

Spawn production is a very sophisticated scientific process; hence most mushroom growers do not produce their own spawn. They buy it from a few, specialized producers who are able to maintain standards and ensure freshness.

Spawn varies in colour depending on age. Very fresh spawn sometimes does not even show any sign of the white mycelium growth of the fungi. If spawn is very densely covered by the white fibrous mycelium growth, this may indicate it is too old. In both of these cases, it will generally still give an acceptable result. Always check spawn for visible signs of non-mushroom growth. Other fungal mycelia may vary in colour or growth form. If it appears that there are two different types of mycelia growing, then the spawn should not be used.

For large-scale production of mushrooms, a large quantity of the specific inoculum is required (often 1-5% of the final mushroom production medium). In mushroom growing technology the

inoculum is known as the "spawn". Spawn is a medium that is impregnated with mycelium made from a pure culture of the chosen mushroom strain. Spawn production is a fermentation process in which the mushroom mycelium is increased after growing through a solid organic matrix under controlled environmental conditions. In almost all cases the organic matrix is sterilized grain, e.g. millet, rye or wheat. The purpose of the grain spawn is to boost the mycelium to a state of vigour such that it would rapidly colonise the selected bulk growing substrate. The grain is an important nutrient support as well as a vehicle for the eventual and even distribution into the growing medium of the mushroom inoculum.

Each individual grain becomes coated with the mycelium and in fact becomes a mycelial capsule. All operations from pure culture isolation through spawn preparation must be conducted under sterile techniques and performed as rapidly as possible to lessen the possibility for contamination to occur.

An extensive technology has been developed throughout the world to ensure the production of high quality mushroom spawn. Many companies now specialize only in the production of mushroom spawn and can ship active spawn to growers when required. Most spawn is now prepared and shipped in autoclavable polypropylene bags with breathing patches. For natural log production of mushrooms the inoculum or spawn can also be in the form of wood chips coated with the specific mushroom strain.

Fig. Sterilization of grains in autoclave

Procedure for production of grain spawn

Equal amount of wheat grains and water (w/w) are boiled for 15 min, and then allowed to soak for another 15 min without heating. The excess water is drained off and the grains are cooled in sieves. The cooled grains are mixed with calcium carbonate and gypsum ($CaSO_4.2H_2O$). The gypsum prevents the grains from sticking together and calcium carbonate is necessary to correct the pH. The prepared grains are filled into half-litre bottles or polypropylene bags (@ 150-200 g/bottle or bag) and sterilized in autoclave at 15-pound pressure (per square inch) for 2 hours. The bottles are inoculated with mycelium and incubated at 22-24 ^{0}C in a dark place.

Materials necessary while transfer of spores or spawning

A clean lab bench (swab with 50% Lysol, 10% bleach, or 70% ethyl alcohol); washed hands (wash with soap and spray with alcohol); clean clothes; and clean air. In addition to these HEPA (high efficiency particulate air) filters are also used, if the transfer is to be taken at high level. With HEPA, the particles of 0.3 micron size are removed from the air with 99.99% efficiency (i.e., only one out of every 10,000 particles of that size could pass the filter). It is worth to mention that typical clean desert air contains 200,000 particles per cubic foot and in urban areas this value is raised tenfold.

SPAWNING

The sowing of spawn in the appropriate substrate is called spawning. The quality as well as quantity of mushroom is depends upon the process of spawning.

Boil grain with equal volume of water for 15-20 minutes

↓

Drain off excess water

↓

Surface dry the grain shade for.a few hours

↓

Mix the grain thoroughly with Calcium sulphate (2%) and Calcium carbonate (0.5%), on dry weight basis of grains

↓

Fill 300-350 g grain mixture in 500 ml conical flasks

↓

Plug the conical flask with cotton plug sterilize in autoclave for two consecutive days at 15 PSI for 30 minutes

↓

Allow the conical flasks to cool down to room temperature

↓

Inoculate conical flasks with culture in sterile condition

↓

Incubate at 20-30 ^{0}C in dark for 3 weeks

↓

Shake Containers periodically for distribution of mycelium, as the mycelium and grains form a compact

Fig. Protocol for button mushroom's spawn production

Methods of spawning

There are different methods of spawning which are given below:

1. **Double layer Spawning**: In this method cultivation tray is half filled with compost. The spawn blocks are plotted on the compost at a uniform spacing. Then the tray is filled completely with compost and after the filling spawn blocks are placed again. The blocks are pressed with the fingertips into the compost and finally tray is covered with polyethylene sheet.

2. **Top Spawning**: In the top spawning, the cultivation tray is filled with compost and at the top, spawn blocks are placed and then a fine layer of compost is spread. Finally a polyethylene sheet is spread as a cover.

3. **Thorough Spawning:** In this method the compost is thoroughly mixed with the spawn and then the trays are filled with this mixture.

4. **Shake-up-Spawning**: In such method, a week after spawning, the compost is taken out from trays and stirred properly in order to mix both spawn and compost. Then the trays are filled again and cased with mud soil.

5. **Active Mycelium Spawning**: This method of spawning was invented in Germany. Here in this case, just before appearing the pinheads, when the mycelium covers the compost. The compost mass of the tray is stirred and used for spawning a freshly filled compost tray.

6. **Spot Spawning:** In the spot spawning, small holes are made on the compost filled in the trays with the help of fingertips then the spawn is placed into them and finally compost is spread over it.

7. **Super Spawning**: In such method trays are filled with pasteurized compost as usual. After two weeks of growth, the compost of all trays are taken out and kept together. The compost of each tray is broken and stirred properly to disturb

it. The fresh stock of compost is taken and mixed with previous mixture and then trays are filled again.

Care and maintenance of crop after spawning

1. Grains of spawn should be separated from each other as thoroughly as possible as the spawn is spread over the surface of the compost.
2. Spawn should be mixed evenly throughout the compost. Conditions should be kept as sterile as possible (Wear clean clothing and footwear, wash your hands before carrying out spawning, tools should be sterilized in, formalin or some other antiseptic which will not damage the mushroom). Do not add spawn to compost while the temperature of the compost is above 30 degrees centigrade. (34 degrees will kill the mycelia).
3. If there is any ammonia present in the compost (i.e. through composting being incomplete), the mycelia are not likely to grow.
4. The optimum temperature for the mycelia to grow is 25^0C.
5. Three or four days following spawning, a cottony growth should appear around the grain, and depending on conditions, the mycelia should have thoroughly grown through the compost after about 2 weeks.
6. Once the mycelia have grown to this point it will generate increased heat in the compost. It is important that the temperature be held down, and this is normally done by ventilation or by a cooling system.

Chapter 7

ENVIRONMENTALLY SAFE PRACTICES

Spawning

The process of spawning is equivalent to the planting of seeds for a field crop. The spawn arrives on the farm in sealed containers that exclude other microorganisms. While spawn is being mixed with mushroom substrate, it is considered good practice to keep the area clean and to close the doors where possible. Cleanliness keeps undesirable fungal spores and bacteria from entering during the mixing procedure.

Immediately after spawning, a grower may choose to place a thin film of polyethylene over the spawned substrate to increase CO_2 concentrations in the tray or bed, help preserve water in the compost, and protect the surface from pests and mushroom pathogens. The film is not generally reusable and should be properly discarded once removed. Good practices call for the film to be placed without delay in the refuse for removal by a properly licensed solid waste disposal company. While awaiting removal, the plastic should be protected from rain to eliminate the potential for runoff of low-level pesticide residues from the plastic surface.

Casing

The fully colonized substrate is covered with a layer of casing material to induce the formation of the mushrooms. This step involves a major input of solid material to the mushroom farming operation.

Recommended practices include the reasonable activities related to good housekeeping. If casing material is stored outside in bulk, it should be treated as other raw materials, and erosion into streams or storm sewers should be prevented. Inside, good housekeeping dictates that any spilled casing material should be squeegeed and picked up. The squeegeed materials are not to be placed with new casing for use on beds or trays because contact with the floor may add impurities to the material.

The water used to moisten the cased substrate may also be used as a means for the application of low levels of minerals or pesticides. In these cases, especially, water should be applied at a rate low enough that the casing would absorb it all.

Multiple small waterings should be used rather than a single large watering to avoid exceeding the absorption rate of the casing.

Thus, there should be essentially no water or pesticide runoff from the casing to the next lower level or onto the floor.

A recommended practice on mushroom farms is to use a dedicated tank and pump system for watering. This system works as follows: the tank is filled only to the volume needed for a specific water application. If there are to be additives for that water application, those powders or solutions are placed in the tank and mixed. All pesticides mixed in the tank must be used according to the pesticide label directions. Mixing must continue during the pumping. It is a recommended practice also that only sufficient mixture be prepared in the tank for a specific room or house, so that there is no leftover chemical requiring storage or disposal. The tank and pump system are useful for plain water application as well as for water with additives because the pump assures a constant pressure at the sprinkler head. In contrast, the well pump connected to a pressure-tank provides a varying pressure, particularly as other valves are opened and closed. Good water quality practices **do not allow** the use of a pesticide injection system that will place chemicals or pesticides directly into fresh water supply pipes.

Fruiting and Crop Management

Pinning is initiated by shocking the mycelia with an introduction of fresh air to lower temperatures and carbon dioxide concentration. Therefore, pinning does not have any environmental effects in terms of odours, water and solids management.

Once the pins are formed and begin to grow larger, the casing needs more water, but the amount of water should not be too much for the casing to absorb. Moreover, applied water may serve as an additive for controlling undesirable bacteria or fungi.

Harvesting Operations

Mushrooms are generally harvested by hand picking. After picking the mushroom, the harvester trims off the ends of the mushroom, often called stumps, and accumulates them in a container for removal from the room or house. Harvesting also generates some over-ripe and misshapen mushrooms that have no commercial value (culls) and a small amount of dislodged casing material.

Fallen harvesting debris should be swept from the floor and placed together in the container specifically designated for stumps.

Stumps may be placed into the heap to become part of future substrate or may be spread thinly on cropland. Stump material should not be left in piles near mushroom houses or elsewhere because it rapidly degrades, attracting flies, sustaining their life cycles, and generating odors.

Workers move from house-to-house or room-to-room according to the harvest schedule. To reduce the transport of mushroom diseases, the workers may wet the bottoms of their footwear on a mat saturated with disinfectant. These mats can be located at the door(s) to each room or house. Also, the workers' tools may be brushed clean and dipped in disinfectant. As with pesticides, spills of these disinfectants must be avoided, and spill control and clean up measures must be in place.

As we know that mushrooms grow in cycles (flushes, breaks, blooms). Between flushes, water is applied to the casing to supply the next mushroom crop. This water application should be performed in the same careful manner as mentioned above under casing.

Post-Harvest Sanitation

After the last picking, the room or house may be prepared for a heat treatment. It is good practice to add live steam to the room or house for the purpose of pasteurizing the substrate and interior surfaces. This practice reduces the spread of disease prior to moving all of the substrate and casing material from that location. Used compost should be removed from the trays or beds and handled in an environmentally safe manner. The room or house may be further disinfected or sanitized after the removal of compost.

Pest Management

In the mushroom crop there are two fly species (Phorid and Sciarid) that feed on mushroom tissue and have the potential to cause economic damage to the mushroom farm. These must be controlled to numbers below the economic threshold level. Integrated Pest Management (IPM) techniques aid the farmers in minimizing the quantity of chemical insecticide needed for fly control. IPM requires that the farmer know the level or number of flies in the room, so a monitoring device is essential. By observing the number of flies at the monitor, the experienced mushroom farmer can determine the flies' time of entry, life cycle, and potential for infestation. This is critical information for the farmer to judge the potential economic impact.

Using the observations at the monitor, the farmer can schedule the timing and dose of insecticides. The IPM technique helps make most efficient use of expensive chemicals.

Good housekeeping and farm sanitation practices help to eliminate breeding locations for all species of flies. As mentioned earlier, the stumps and overripe mushrooms should be removed from the harvest areas to reduce that source of growth media for flies.

Various species of fungi, bacteria, and viruses attack mushrooms. The farmer should use good sanitation practices as the first line of defense against these pathogens. In the event that an infestation occurs, and economic impact will be felt if the disease were to go unchecked, i.e., the economic threshold is exceeded, then chemical or biological control methods would be needed.

It is important to have as much information about the disease as possible so that pesticide use is most effective, not excessive. From both an environmental and economic viewpoint, minimal use of pesticides is best. The farmer must apply pesticides according to label directions when needed to control a specific problem.

Trash Management

General trash and refuse items on a mushroom farm include typical household items such as newspaper, magazines, printout paper, lunch containers, beverage cans, etc. Unique to mushroom farms are: the plastic film used to cover the substrate during spawning; wood (lumber) not strong enough to stay in service for trays, beds, or shelves; pesticide containers; and broken harvest containers. Spawn and casing packaging should be recycled or added to the general trash. The mushroom farmer should check that a properly licensed waste disposal company removes the waste materials and should assure that the wastes are properly disposed of at an appropriate disposal site.

Chapter 8

CULTIVATION OF WHITE BUTTON MUSHROOM

Agaricus bisporus (Lge.) Sing.

The most commonly cultivated mushroom in the world is white button mushroom. Its several strains vary in growth habit, colour, yielding ability, and other characteristics. The most common variation occurs in colour, and three groups *viz.* whites, creams, and browns.

Fig. Fruit body of *Agaricus bisporus*

Many historians consider, *Agaricus brunescence* (=*A. bisporus*) as characterless cultivated variety of field mushroom (*A. campestris*) becoming the most common mushroom grown and sold in the world. They are strictly cultivated in rich compost in special mushroom houses where heat and humidity are carefully controlled. The white button mushroom and its various forms (Crimini, Portobello, and Portobellini) grow best in a moist, cool, well-ventilated place, on either synthetic or natural compost. These factors are either available in nature or in controlled condition, in both ways; the production can be achieved as usual anywhere irrespective of place. Most of the factors necessary for the development of this Mushroom are temperature, humidity, ventilation and type of spawn, etc.

REQUIREMENTS FOR GROWING BUTTON MUSHROOM

Temperature

The growth of mycelium of *Agaricus* spp. is optimum at 22 to 25 ^{0}C temperature. However, the ideal temperature for the growth of pinheads is always 14 to 18 ^{0}C. Fluctuations in the temperature certainly harm the production and quality of the mushroom. Low temperature retards the growth of the crop, whereas high temperature attracts the development of moulds and bacteria.

White Button Mushroom can be grown up to an elevation of 2000 meters MSL (mean sea level) and at least five crops can be raised in a year, if the environmental conditions are met appropriately.

Humidity

The growing room requires special conditions. In addition to cooling requirements (60 to 63° F), the humidity must be maintained at about 95% (or even higher in some situations) with just enough air movement to avoid pockets of stagnant air.

Humidity varies according to available temperature. Maximum humidity in the atmosphere is required for the growth of Button Mushroom. If, atmosphere becomes dry, sprinkling water can increase the humidity; however, it should be adopted regularly as spraying of water after composting is prohibited.

One way to humidify a growing chamber is to bubble a stream of air (using an aquarium air pump and porous stone) through water slightly warmer (using an aquarium heater) than the air to be humidified. Now a day, various humidifiers are also available in the market, which can be used for the purpose. Aquariums, plastic tents, ice chests with cellophane windows, etc., are potential humid chambers used for growing.

Excessive air movement, even with high relative humidity, will dry the surface of a substrate and can damage delicate pinheads.

Creating a very humid environment for mushroom production is perhaps the limiting factor in the cultivation.

Ventilation

The growers of Mushroom are recommended for reducing the Carbon dioxide (CO_2) level in the growth room, as the CO_2 is harmful for the growth of Mushroom. It is required to maintain a CO_2 level of not more than 0.15% of total volume of air. This may be achieved by maintaining proper aeration/ventilation changing 4 to 6 air changes/ hour or introducing 10 cubic foot fresh air/ square foot area, in the cultivation room. It is worth to mention that every degree centigrade increase in the temperature above 16 ^{0}C, increases the production of CO_2 by 25%.

Good Spawn

The seed of mushroom having no spores is called spawn. The spawn for production should be taken from an authentic and strain specific source. For detail information of spawn refer chapter "Spawn".

MAJOR STEPS OF CULTIVATION

The process of mushroom cultivation takes about four months beginning with the preparation of the compost to the harvest. The basic steps in the growth cycle of mushrooms are:

1. Compost and compost making (composting)
2. Sowing of Spawn (Spawning)
3. Layering with Soil (Casing)
4. Picking the fruiting bodies (Harvesting)

1. COMPOST AND COMPOST MAKING

The white-button mushroom is grown on a selected substrate, which provides an adequate level of nutrients to support the crop.

The substrate, which is used for growing mushroom, is called **compost**. The process of making compost is called **composting.**

Ingredients of compost and their functions

As we know that mushrooms have no chlorophyll, they require nutrients from organic matter in their growing medium, if grown. The medium is scientifically formulated using various materials such as:

1. **Base materials** include wheat straw, maize cobs and other similar cellulosic plant wastes with or without horse-manure, to provide celluloses, hemicelluloses and lignin in bulk materials.
2. **Supplements:** These activate fermentation.

i) **Animal dung:** This includes horse and poultry manure having nitrogen content from 1-5%. Cow manure is not suitable for compost. However, one-third cattle dung mixed with two-third amount of horse dung may serve the purpose. The unsuitability of cattle manure in composting is due to high variability of the raw material used for the cattle feed. The poultry manure is used to replace part of the organic nitrogen supplement and to give good yield as it adds a significant amount of fine dry matter to the compost which helps to counteract the strawiness of horse manure and improve the texture of the compost.

ii) **Carbohydrate nutrients:** Such nutrients are added from molasses, potato wastes, apple & grape pumice and malt sprouts, etc. These materials are required to balance the C : N ratio and establishing bacterial flora.

iii) **Concentration meals** include wheat or rice bran, the seed meals of cotton, soybean, castor, sunflower and linseed. These materials supply both nitrogen and carbohydrates. Nitrogen content may vary from 3-12% depending upon the source.

iv) **Nitrogen fertilizers** like ammonium sulphate, calcium ammonium nitrate and urea are rapidly released for the quick growth of microbial population. Addition of nitrogenous fertilizers such as ammonium sulphate should be done according to formulae as it delays the decomposition in the pile.

v) **Materials to correct mineral deficiencies**: In addition to carbon and nitrogen, Muriate of potash and calcium superphosphate are required by *A. bisporus*.

vi) **Materials to correct greasiness:** Gypsum and calcium carbonate serve to precipitate suspended colloidal materials and neutralize greasiness.

Formulations

Formulations are based on cost and availability of raw materials in the particular region. To initiate the composting and minimize the loss of dry matter, generally 1.7-1.8 percent nitrogen is maintained. The main objective of computing a formulation is to achieve a balance between carbon and nitrogen compounds. The C:N ratio is adjusted to 25-30:1, which comes down to 16:1 after composting. Gypsum is added @ 2% of dry ingredients, which adhere to straw and permeate air in the pile more readily, as it is essential in the process of composting. In nutshell, a formulation must be prepared in such a way to have 1.7-1.8 % nitrogen, 1.2-1.5 % phosphorus, 2.0-2.3 % potash, 1.5-3.0 % calcium and 0.4-0.5 % magnesium.

Natural Compost

It is prepared from horse dung. In the stable plenty of straw is used as litter for the horses'; dung, litter, and urine are mixed together to form stable manure. It is desirable to add one-third of the weight of wheat straw to the horse dung in order to make good stable manure. If litter is not used in the stable, necessary straw, dung and urine should be mixed. The horse dung must be fresh and must not have been exposed to rain. The mixture should be piled in a way of heap having 1 meter height. The compost pile should not be exposed to the sun and shower. After 3-4 days the fermentation starts to increase and also production of ammonia gas starts, following the temperature increase. Pile should be broken and re-piled to provide proper aeration/ventilation. The process is repeated 4-5 times at an interval of 4-5 days. At the last operations (second and third turning) some amount of gypsum is added at the rate of 25 kg/tones of compost. At the final turning nematicide should be mixed in manure.

Synthetic Compost

Mushroom cultivation from the beginning was grown on the substrate of natural compost however, ingredient are more in lesser amount. A formula suggested by Kapoor and Bahl, 1982 is adequate for production of Mushroom of all kinds, which include:

1.	Wheat or paddy straw (8-10 cm cutting)	250 kg.
2.	Wheat bran	25 kg.
3.	Ammonium sulphate /Calcium Ammonium nitrate	4 kg.
4.	Urea	3 kg.
5.	Gypsum	20 kg.

Method of composting

There are two methods for composting, the long and the short methods.

Long method of Composting (LMC)

This method of composting was first advocated in India by Mantel *et al.* in 1972. Compost prepared by this method takes more time

(28 days) and give low yield as it is prone to attack by many pests and diseases. This compost generally converts into approximately 8-12% mushroom yield (100 kg compost yields about 8-12 kg of mushroom). The compost is prepared on clean cemented platform. The size of platform depends upon the several factors. Generally a medium sized farm producing around 20 tonnes of compost in one go, require a platform size of 60′ x 40′.

Before starting the composting, the compost platform and nearby area should be sanitized with 2% formaldehyde solution. This operation is necessary in LMC as compost is not sterilized in this method.

Procedure for LMC

i) **Wetting the straw:** The straw is spread thinly over cemented platform. The straw wetting is done by sprinkling water. Several turnings are given for uniform distribution of water. Wheat straw included in composting is a very hardy material as well as coated with wax and does not absorb water quickly. Excess amount of water, which generally leaches down, is collected in a pit built nearby to yard. This collected water may be recycled in the process of watering. Wetting is continued till the materials have gained 70 percent of moisture.

ii) **Mixing and heaping:** After the straw is wetted, the supplements excluding the gypsum are uniformly scattered over the straw and mixed and the mixture is finally stacked in rectangular form having 1 x 1 meter height and width. The length depends on the quality of compost material. The heap is compressed by applying light pressure with the help of wooden board.

iii) Detailed schedule of composting

Schedule	Work to be done	Remarks
Day zero	Wet, mix and stack the pile	Nitrogen and carbohydrates are supplemented.
6th day	First turning	
10th day	Second turning	
14th day	Third turning	Half amount of gypsum is added
18th day	Fourth turning	Another half amount of gypsum is added
22nd day	Fifth turning	Spraying with 40 ml of nemagon
26th day	Final turning	Application of 10 ml malathion in 5 litres of water.

During first turning itself it can be observed that the mixture which was yellowish in colour, has turned to dark yellow to light brown in colour and shrink noticeably. There is a large production of ammonia which can be observed with the foul smell it leaves. Ammonia is produced under anaerobic conditions by the breakdown of carbohydrates and proteins. In addition to ammonia, a large amount of CO_2 is also produced. After final turning the ammonia smell should be observed, if there is ammonia smell, one more turning should be given to wave off ammonia smell from compost completely.

Modification of LMC

The above method of composting generally takes around 28 days. A little modification in the method has been suggested by Dhar and Kapoor (1989) which completes composting in 16 days. As the schedule compresses (Zero day stacking; 5th day first turning; 8th day 2nd turning; 10th day 3rd turning and addition of gypsum; 12th day 4th turning; 14th day 5th turning and addition of insecticides;

on 16th day break opening the pile) to 16 days, the chances of ammonia production is quite high, which needs to be removed. Solar pasteurization of compost prepared in 16 days is also advocated as it attracts most of the pests. For solar pasteurization the compost is spread (2.5 cm thick) over the cemented floor in intensive sunlight for 8 hrs and covered with polythene sheet. Grewal (1989) has suggested for addition of dried leaves of *Azadiracta indica, Cannabis sativa, Eucalyptus tereticornis* and *Ricinus communis* @ 3 kg/ 100 kg of dry wheat straw for controlling pests in composting.

Short Method of Composting (SMC)

The idea of SMC traced back to 1915 when American growers noticed that if prepared compost is kept in shelves in growing rooms and subjected with high temperature (60 ^{0}C) for some time gives the higher and consistent yield of *A. bisporus*. This process, in beginning, was termed as "sweating" of compost. Lambart in 1941 analyzed the whole process of composting and come to the conclusion that the "sweating out" process should be made as an integral part of composting. Finally the short method of composting was developed by Sinden and Hauser in 1950 with following formula, which consists of two phases.

1.	Chopped wheat straw	1000 kg
2.	Chicken manure	400 kg
3.	Brewer's grain	72 kg
4.	Urea	14.5 kg
5.	Gypsum	30 kg

Procedure for SMC

The procedure for phase I is similar to the initial stages of the LMC except, that turnings are given sooner, the first on day 3, the second on day 6 and the third day on 9 or 10 when gypsum is added. The compost is now ready for the phase II or the peak heating stage. This is recognised as the microbial-composting stage and is an integral part of the total composting process. The phase II of composting is comprised of pasteurization of the compost in order to kill microbes. It is done by heating the compost, at 60^{0}C for a brief period through warm air.

1. Phase I (Pre-Conditioning)

Like LMC, The raw materials are mixed, moistened and allowed to undergo some initial decomposition in a large heap. Leached water is regularly sprayed over the raw materials. A mixture of manure and straw usually makes up the bulk of the compost. It should be ensured that material should be kept moist but should never be over wet during whole stage. After 2 days the stack is broken, water is added to the dry portions and again an aerobic stack is made. To ensure fully aerobic condition at least 1-15 M^3 of air/ tonne of wet compost/hr must pass. For this purpose required capacity of small blowers are provided at one end of the yard.

In this phase like LMC the ingredients are allowed to ferment under controlled condition and partially decomposition is achieved. After a successful composting, the centre of the heap becomes hot (80 ^{0}C) in a couple of days resulting in to the charring of ingredients giving distinct brown colour to the mass.

Phase I composting lasts from 7 to 14 days, depending on the nature of the material at the start and its characteristics at each turn. There is a strong ammonia odour associated with composting, which is usually complemented by a sweet, moldy smell. When compost temperatures are 70 ^{0}C or higher, simultaneously ammonia is present, the chemical changes occur which result in a food rather exclusively used by the mushrooms. As a by-product of the chemical changes, heat is released and the compost temperatures increase. Temperatures in the compost can reach 75 to 78 ^{0}C during the second and third turnings when a desirable level of biological and chemical activity is occurring. At the end of Phase I the compost should:

a) Have a chocolate brown colour;

b) Have soft, pliable straws,

c) Have moisture content of from 68 to 74 percent;

d) Have pH between 8-8.5

e) Have a strong smell of ammonia (800-1000 ppm).

When the moisture, temperature, colour, and odour described have been reached, Phase I composting is completed.

2. Phase II (Pasteurization and Conditioning)

This phase is also called finishing of compost. There are two major purposes to Phase II composting. Pasteurization is necessary to kill any insects, nematodes, pest fungi, or other pests that may be present in the compost. Secondly, it is necessary to remove the ammonia, which formed during Phase I composting. Ammonia at the end of Phase II in a concentration higher than 0.07 percent is often lethal to mushroom spawn growth, thus it must be removed; generally, a person can smell ammonia when the concentration is above 0.10 percent.

Phase II composting can be viewed as a controlled, temperature-dependent, ecological process using air to maintain the compost in a temperature range best suited for the de-ammonifying organisms to grow and reproduce. The growth of these thermophilic (heat-loving) organisms depends on the availability of usable carbohydrates and nitrogen, some of the nitrogen in the form of ammonia.

Phase II process of composting may be divided into two categories, *viz.* Pasteurization of compost and Conditioning.

i) Pasteurization of compost

Pasteurization of compost is carried out in two ways. In one case filled trays, may directly be put in the peak heating room (tray pasteurization). Secondly the bulk pasteurization may be carried out in tunnels, especially made for the purpose.

a) Tray Pasteurization

Trays evenly filled with compost are stacked in a pre-conditioned room quickly to avoid heat loss. Sufficient space is ensured between the trays while stacking to allow free movement of the air. All doors and ventilators should be closed; if possible insulated room is selected.

After the proper stacking of compost filled trays, the live steam is introduced to raise the temperature of compost up to 52-54 ^{0}C; this temperature is maintained for 2-4 days. After this period, the temperature of the room is further raised between 58 to 60 ^{0}C and maintained for 4 hours. The temperature above 60 ^{0}C is harmful as it may kill all kinds of organism including thermophilic fungi, which are very essential for governing phase II of composting. The steam supply is interrupted and fresh air is introduced slowly to lower the temperature by 2-4 ^{0}C during next 24 hours. The temperature is still lowered and maintained to 52-54 ^{0}C for 4 days. During this period, the ammonia gas disappears. After this stage, the temperature of compost is further lowered to 24 ^{0}C, the usual temperature necessary for spawning.

b) Bulk Pasteurization

In this process compost is pasteurized in bulk inside a tunnel that is specially built room known as tunnel. The compost is filled in the bulk chamber up to the height of 2-2.2 m; in such a way that one square meter of space occupies approximately 1 tonne of compost. Several temperature sensors are placed at different points of the tunnel to measure the temperature. One sensor is paced below the plenum in the ventilation duct below the grated floor; two or three inside the compost mass and two above the compost for air temperature.

Immediately after filling, all the doors are closed and blower is switched on to bring uniform temperature of 45 ^{0}C throughout the compost. As the temperature reaches to 45 ^{0}C, the introduction of air is stopped. Now the compost temperature starts rising, @ 1.2 ^{0}C/hour reaching the required temperature of 57-58 ^{0}C for pasteurization in 10-12 hours. Simultaneously some amount of steam may also be passed to the compost chamber. This temperature is maintained till the no detectable limit of ammonia is achieved. The temperature is brought down to 25-30 ^{0}C by introduction of fresh air in the tunnel.

ii) Post pasteurization conditioning

The pre-pasteurization conditioning implies to achieve the maximum population of thermophillic flora, which are generally lost during the process of pasteurization, hence it necessitates the post pasteurization conditioning. This conditioning again builds the lost thermophillic organisms. During the process temperature is maintained at 45-52 ^{0}C and enough oxygen is supplied to maintain at least 10% oxygen.

Precautions while composting

It is necessary to chop the straw in order to facilitate the retention of heat and moisture both during composting period and during process of sweating. Chopped straw is also easier to handle, as long straw is laborious to mix. Moreover, composting period is dependant upon the material of compost.

The compost heap should be smaller in size, as small size piles receive good aeration and exposure to atmosphere. CO_2 concentration in the pile is dependant on the height and width of the pile, therefore smaller the pile, lower the CO_2 level.

While composting moisture level should be maintained at par because if moisture is low during composting the bacterial decomposition is arrested, nitrogen escapes as ammonia and other volatile forms of nitrogen. High level of moisture brings anaerobic condition thereby lowering the speed of decomposition process. If the compost is wet more that 70% automatically the supply of oxygen will be arrested. Low oxygen level causes the high level of carbon dioxide in the compost.

The interval of turning is crucial in the process of composting, which is governed by the nature of composting material. If the turning is delayed there will a development of anaerobic condition, slowing down the decomposition of compost.

Qualities of ready compost

- The compost heap appears as light brown in colour,
- The straw of the compost material should have broken, but not in very small pieces,
- The heap should have lost the smell of ammonia,
- It should have the aroma of mushroom,
- Upon squeezing it should not leave a water, rather moisture,

SPAWNING (planting spores)

Adding spawn into the compost is called spawning. The compost is placed in beds or trays, under cover, and spawn is added. Conditions are controlled to achieve optimum growth of the spawn so that it can fill the compost with mycelia over a stipulated period.

Spawn is the propagating material for mushrooms. It contains spores mixed with spawning materials such as manure, tobacco stems, or grain. Spawn is most often purchased from the authentic sources.

Spawn is planted either by the broadcast method or by planting in small pieces 1½ inch deep at 10-inch intervals. In white mushrooms, spawn is added at a rate of 2.5% or more of the dry weight of the substrate.

Although spawning and spawn run can be accomplished in any clean room at cardinal room temperature with fresh air, sterile techniques, which are absolutely necessary for successful cultivation of white button mushrooms.

Spawn Run: Depending on the growing condition, the substrate is usually fully captured by the mycelium within 2 to 6 weeks. Spawn run temperatures should be 22-25 ^{0} C.

CASING

After spawning the surface of compost is covered with a layer of paddled clay, the operation is called casing. It involves placing a layer on top of the compost, to bring about environmental conditions encouraging the formation of fruiting bodies. Casing material is normally applied about 2-3 weeks after the spawn has fully run through the compost. A uniform thickness is maintained as water sprinkling ensures the proper absorption of water in the compost. Besides this pinheads can **pearg** the thin layer very easily. Therefore, a uniform thickness of 2.5–4.0 cm should be maintained.

The full run of spawn is observed as visible masses of gray white fibrous mycelia appearing through the compost. One should ensure the even distribution of layer making it one inch throughout the compost tray or bed.

For the first time French farmers understood the utility of casing and its use after test experiment on button mushroom. French scientists for the first time used limestone ($CaCO_3$) as casing material.

The mycelia, which grow throughout the compost, following spawning, are the vegetative growth stage of the mushroom. Actual edible mushrooms are the fruiting stage of growth, and as such, they do not begin to form until something happens to stimulate a change from vegetative growth to fruiting body formation. Casing stimulates the required change and mushroom fruiting body through pinheads starts to appear. We still do not fully understand why casing causes this change to occur.

Advantages of casing

- Casing helps in changing the active phase into reproductive phased, i.e. the state of sprouting fruit bodies that bears spores to complete the life cycle,
- Casing makes proper fruit size,
- Casing preserves moisture of the compost for a long time. The moisture is indispensable for growing quality product in good quantities.
- Casing preserves nutrient loss as well as maintains the temperature of the bed, as it repels the heat from its surface.
- Casing absorbs water and passes slowly to compost underneath after spraying of water.

Quality of good casing material

- Casing material should have an open texture, should not repel water (as do some organic materials) when dry, and have a pH between 6.5 and 8.
- Casing material should normally be as wet as possible when applied to the beds and is usually applied to a depth of 25 to 50cm over the top of the compost.
- It should have good water holding capacity, also it should absorb the water quickly and release slowly.
- The casing material should be porous in order to favour the aeration.
- Material should be free from pests including diseases.

Material for casing

There are many different materials, which can be used: pulped paper, compost, cow dung, old/used mushroom compost, peat and sphagnum mosses, or mixtures of any of these. The casing material usually preferred should have high organic material, and free from pests and diseases (ideally sterilized). The material can be sterilized by drenching in formaldehyde solution or by pasteurization at 60 ^{0}C for 3 hours. Now days a mixture of two or more materials is used for casing. The following are some important formulae used for preparation of casing material:

- Good manure and sandy loam soil at the ratio of 3:1
- Clay and sand (1:1)
- Farmyard manure and grit (4:1)
- Farmyard manure and loam (1:1)
- Compost, sand and slaked lime (4:1:1)
- Well rotten cow dung, mixed with light soil (3:1)
- Soil and sand manures (1:1)
- Farm yard manure and gravel (4:1)
- Farm yard manure and loam (1:1)
- Soil peat mixture (2:1)
- Spent compost, sand and slaked lime (4:1:1)
- Nematicides mixture

A simple formula for casing material, which has been innovated include, mixing fine texture sandy loam with pure leaf **mould** in the ration of 1:1. Before mixing, leaves should be sieved through a sieve of 2 mm mesh. Soak the mixture in formalin solution (2%) and keep covered for 48 hours. After this period allow it to escape the formalin for about a week. Before applying as casing material, add water to make a puddle of pliable consistency. One should ensure that it should be free from spore of diseases and pests.

Pasteurization of casing material

Casing material is pasteurized in order to make it free from germs. This may be achieved either by applying heat or using chemicals. Heat in pasteurization of casing material may be applied in way of steam through perforated pipes. The temperature of steam should be kept at 60 ^{0}C and maintained at least for 5 hours. It is worth to mention that the term 'sterilization' is intentionally being avoided, as it is pre-advised that sterilization of casing material under pressure kills several beneficial microorganisms leaving material more vulnerable to diseases and pests.

There are some chemicals of sterilant nature like formalin, chloropicrin, methyl bromide, vapam etc. are also use for the purpose. Out of them formalin is widely used at the rate of 500 ml mixed in 10 Liter of water for treating one cubic meter of soil to be used for casing. The casing material is spread in a thin layer on a sheet of polyethylene sheet. Upon this thin layer of material, spray formalin solution and pile the treated material and finally cover the material with polyethylene sheet, for 48 hours. After week time formalin gas gets evaporated and then the pasteurized material will be ready for casing purpose.

Growth and development of mushroom crop

Generally from 15-20 days after casing the fruit bodies begin to rise. At this times water should be sprinkled. Humidity of room should be 70-80 %, or the compost of the tray will tend to dry resulting in the bad growth of mycelium. The ventilation within the room should be normal or the crop. If heat and humidity be increased the incidence of the pest and disease will occur. For this reason the temperature should at 14-16 ^{0}C. No bad smell should be there in the room. If the ventilation is good no foul smell can be experienced. In the dry weather the soil of casing will run dry. To prevent this condition damp the floor by sprinkling water.

These protrusions are called as **pinning**. The pins continue to grow, becoming the mushroom caps, which are actually the fruit of the plant, just as a tomato is the fruit of a tomato plant. During **pinning** temperatures are reduced for 2 days to 2 weeks and CO_2 level is lowered by introducing fresh air. While **fruiting,** colonized trays of substrate should be placed in a growing room maintained at cool temperatures (63° F) and high relative humidity (85-95%), so that good crop is expected. It takes 17 to 25 days to produce mature mushrooms after the casing material is applied. Size is no indication of maturity in mushrooms. Perfectly ripe ones vary from small buttons to large caps.

HARVESTING

After several weeks, the first crop is picked. Further crops occur every week or so (these are called flushes). The fruit bodies rise collectively, i.e. in **flocks** and in successive **batches**. It takes a weeks time to attain the size of large button. The time gap between the first and the second batch is usually 10 days. After a month or two, the flushes decline also the size of mushroom will comparatively be small. At this moment the compost should be discarded and the process started again.

Fig. Mushroom ready to be picked up

Each mature mushroom is between 1 and 2 inches in diameter. Harvesting may be done daily or every few days, depending on sprouting. Generally when veil is loose and free from the cap fruit body is ready for picking at this time fruit body remains tight. Fruit bodies are picked up by twisting the stem gently off the soil. It is said that fruit bodies should be picked in the early hours of the day by light twisting so that the broken stump (stub) is not left over in the trays and the small developing mushrooms are not disturbed.

For the purpose of canning harvesting should be done when fruit bodies are about 3 cm size. The base of the stalk with the soil clinging to it should be clipped off.

After the crop, house is emptied and steam-sterilized before the process begins again. The remaining compost is recycled for potting soil. The harvested mushrooms are set in carts, refrigerated and then packaged and shipped quickly to supermarkets, food processors and restaurants. The entire process from the time the farmer starts preparing the compost until the mushrooms are harvested and shipped to market takes about four months.

STORAGE

Mushrooms are readily perishable commodities and start deteriorating immediately after harvest. Due to the presence of enzymes like phenol oxidases, they develop a brown discolouration on surface and become soft in due course of time. The rate of respiration in harvested mushrooms is often high, compared to other perishable horticultural crops, resulting is shorter shelf life. Therefore, immediately after the harvest, there is need of storage, preservation and processing of mushroom, if not utilized or sold afresh in the market.

The self-life of fresh mushrooms may be extended by refrigeration at 1-4 ^{0}C. Low temperature retards the growth of microorganisms, reduces the rate of post harvest metabolic activities of the mushroom tissues and minimizes moisture loss. Detailed information on preservation and processing are available in chapter "post harvest technology of mushrooms".

CULTVATION OF *AGARICUS BISPORUS* AT A GLANCE

Cultivation of mushroom requires following steps

1. Preparation of compost
2. Filling of trays beds with compost
3. Spawning of beds
4. Casing
5. Watering of beds
6. Harvesting of mushrooms
7. Storage

Materials required:

- Mushroom house either kucha or pucca
- Cement concrete platform of 14x7 meter size for preparing compost
- Tray beds made up of either pinus or deodar wood of 100x50x20 cm size

Compost ingredients

1.	Chopped wheat straw (3-6 cm long chips)	300.00 Kg
2.	Wheat bran	15.00 Kg
3.	Calcium ammonium nitrate (or Ammonium sulfate)	6.00 Kg
4.	Super phosphate	7.50 Kg
5.	Urea	2.40 Kg
6.	Potassium sulphate	3.00 Kg
7.	Gypsum	30.00 Kg
8.	Saw dust	10-12.00 Kg

- Spawn @ 2.25% of substrate
- Casing soil
- Sprayer with a fine nozzle
- Baskets for collecting mushrooms
- Wooden mould
- Wooden board (25x12 cm) with handle for compressing the compost
- Newspaper sheets dipped in formalin
- Lime
- pH meter

Procedure

A. Preparation of compost

1. Wet the saw dust by spraying water; mix other ingredients except wheat straw

2. Spread the wheat straw over the cemented floor on the following day and sprinkle with water
3. Spread pre-mixed constituents over the wheat straw surface and mix thoroughly
4. Stack the mixture into a pile of 1.30 meters wide and 1.30 meters high, using wooden mould
5. **First turning** 6th day, add fertilizers; wheat bran and spray water
6. **Second turning** 10th day, add gypsum and spray water
7. **Third to seventh turning:** after every 3rd day and up to 25th day
8. Observe for ammonia smell, if there is smell of ammonia allow one turning in order to smell off ammonia

B. Filling of tray beds

1. Spread the prepared compost on the platform
2. Mix 3 kg calcium carbonate to it
3. Fill the compost in all the corners and edges of tray
4. Compress fairly the compost on the tray using a wooden board leaving one cm of clear space on the top of the tray

C. Spawning

1. Perform the spawning by spreading the spawn on tray beds; mix with fore-fingers and fill the rest of the tray with compost
2. Cover the tray with newspaper sheets
3. Sprinkle water on newspaper sheets in order to provide humidity
4. Stack the inoculated trays vertically one over the other depending upon the height of the room
5. Continue water spraying twice a day or less depending upon the available humidity in the atmosphere throughout the spawn running and cropping period
6. Maintain temperature of room 25 ± 2 0C for 12-15 days for running period of spawn

7. Observe the white cottony growth over the compost tray and colour of compost in the tray changes dark to brown

D. Casing

1. **Prepare** the casing material by mixing spent compost, sand and slaked lime in the ratio of 4:1:1
2. Treat the mixture with nemagon @5 ml/cubic meter of casing material
3. Leave the material in a pile of 1.2x1.0 meter size under a tree shade for a period of one year, turning it every 4 months
4. Sieve the spent compost, which turns into the black soil in one year
5. **Observe** casing material turns black in colour and pH between 8.0-8.5 which is most suitable for mushroom yield (***if necessary adjust pH by adding lime or free stones to the casing material***)
6. **Sterilize** the casing material before use by chemicals like formalin, chloropicrin, methyl bromide, vapam (***alternatively it may be done by providing steam from a boiler through perforated pipes of raising temperature to 70-75 ^{0}C***)
7. **Casing the beds** is done by removing the newspaper sheets from the spawned beds after 3 weeks; pressing the compost with the help of wooden board and covering the bed with 2-2.5 cm thick layer of sterilized casing material.

E. Watering of beds

1. Spray beds over the casing soil with a fine nozzle of a sprayer to maintain RH between 70 to 80 per cent.

Observe the beds for mushroom crops which is expected after 5-20 days of casing

F. Harvesting of Mushrooms

Cap still being tight to its stalk is the right stage to harvest the mushrooms:

1. Harvest mushrooms by holding between fore-fingers slightly pressing and twisting

2. Soil particle and mycelium strands clinging to based are cut with a sharp knife
3. Collect mushroom in the specially designed wooden baskets

F. Storage

1. Store the mushrooms at 4 ^{0}C in a refrigerator for a few days to avoid the quality deterioration

2. G. Precautions

1. Always prepare compost over 300 kg of quantity of wheat-straw, as the heat generated in smaller pile is not sufficient for fermentation
2. Compost pile should not become hard pressed to avoid anaerobic conditions as decomposition takes place in aerobic conditions
3. If ammonia smell is noticeable more turnings should be done
4. Water should never be poured over the spawn to avoid the failure of the crop
5. At least one meter gap should be kept between last tray and ceiling

Chapter 9

CULTIVATION OF OYSTER (*Pleurotus* spp.)

Oyster mushroom is also called **Dhingri,** tree oyster, phoenix, wood fungus, or sovereign **mushroom.** *Pleurotus sajor-caju* is widely used for growing. There are several other species are now available for cultivation. These are *P. sajor-caju, P. florida* (probably a variant of *P. ostreatus*), *P. sapidus, P. eryngii, P. columbinus, P. cornucopiae,* and *P. abellatus.* Oyster mushrooms are cultivated and grow well on rotted wood in clusters.

The cultivation of *Pleurotus sajor-caju* spp. is more easier than that of other spp. and also the production is quite higher. The fruit bodies are beautiful hence they are attractive. The nutritive value is also good but the taste is not so delicious as in white button Mushroom and paddy straw Mushroom. Therefore, for this reason this oyster Mushroom is less popular than white button Mushroom or paddy straw Mushroom.

Oyster mushrooms are wild varieties of *Pleurotus* spp. that have become easy to cultivate and therefore are more widely available (and less expensive) than other specialty mushrooms. They come in a variety of colors including off-white, pink, yellow, and gray brown. Oyster mushrooms grow in tight clusters and have a dense, chewy texture. They are delicately flavorful when cooked, and are often used in less assertive tasting dishes. They are also used in the dried form if not consumed fresh.

Pleurotus spp. is one of the choice edible mushrooms, cultivated in the subtropical and temperate zones. In Europe it is known as the oyster mushroom (*P. ostreatus*), in China it is called the balone mushroom (*P. abalonus* or *P. cystidiosus*).

The *pleurotus sajor caju* can be easily cultivated in a thatched hut under normal conditions of temperature (21-28 ^{0}C) and humidity

(55-70%) for a period of 6 to 8 months, in a year. It can also be cultivated in summer months, when provided suitably the extra humidity.

The different species of *Pleurotus* are suited to grow within a temperature range of 15 to 30 ^{0}C. *P. sajor-caju* is tolerant of a tropical temperature of 28-30 ^{0}C although, it fruits faster and produces larger mushrooms at 25 ^{0}C during the cooler months of the year or in the highlands of the tropics. This is the species now popularly grown in the tropical Southeast Asian countries, including India. *P. abalonus* prefers lower temperatures of 22-24 ^{0}C and is most popular among the Chinese. *P. ostreatus* is the so-called low-temperature oyster mushroom, fruiting mostly at 12-20 ^{0}C. This species is more suited to the temperate climates of Europe and the United States, although many growers in the USA are also producing *P. sajor-caju*.

Nutrition in *Pleurotus sajor caju*

Mushrooms, also called 'white vegetables' or 'boneless vegetarian meat' contain an ample amounts of proteins, vitamins, fibers and medicines. Mushroom contains protein, which is higher than those of vegetables and fruits and is of superior quality. It is considered ideal for patients of hypertension and diabetics. According to Janadaik and Kapoor, 1975, the *Pleurotus sajor-caju* contains:

Ascorbic acid	0.06%
Fat	2.26%
Protein	47.93%
Reducing Sugar	0.285%
Starch	0.0120%

The fruit bodies of *Pleurotus* species have been discovered to have definite nutritive and medicinal values. They are a good source of non-starchy carbohydrates, dietary fiber (that can help in reducing the plasma cholesterol), most of the essential amino acids, minerals and vitamins of B group, and folic acid (necessary to counteract pernicious anaemia) in particular. Considering the essential amino acid index, biological value, in vitro digestibility, nutritional index, and protein score, *Pleurotus* species fall between high grade vegetables and low grade meats. Fractions of water-soluble polysaccharides are reported to possess anti-tumor activity. The physiological processes such as changes in water content, respiratory rate, texture, color, and activities of enzymes like proteases and polyphenol oxidases during the after-harvest life are delineated.

Spawn for cultivation of oyster mushroom

The spawn of mushroom (ramified mycelium) is used as seeding material for the cultivation of mushrooms. Dhingri mushrooms can be grown directly using the master spawn or the multiplied spawn. One bottle (500 ml capacity) of spawn after multiplication as a substrate spawn can produce about 4 to 6 kg of fresh mushroom.

Substrate for oyster mushroom

Dhingri mushroom can easily be cultivated in rural areas to supplement farmers' income, by utilizing the locally available materials.

Like the other mushrooms, Pleurotus spp. can be grown on various agricultural waste materials, with the use of different technologies. They grow well on different types of lignocellulosic materials, converting the materials into digestible and protein-rich substances suitable for animal feeds.

Pleurotus spp. may be produced in the tropics on a mixture of sawdust and rice bran, rice straw and rice bran, sawdust and other combinations of tropical wastes. Other wastes such as corncobs, cotton waste, sugarcane bagasse and laves, corn leaves, grasses, rice hulls, and water hyacinth leaves are also good substrates for growing this mushroom. The substrates used in each region depend upon the availability of agricultural wastes.

Formulation for compost

This formula is specially used for nylon netting rope cot method of cultivation of oyster mushroom. Several ingredients are added while making compost for cultivation of oyster mushrooms, these include:

Straw chips (1-2.5 cm long)	15 kg.
Straw chips (1-2.5 cm long)	15 kg.
Urea	325 gm.
Super phosphate	800 gm.
Sulphate of potash	300-600 gm.

In the formula, urea, superphosphate and sulphate of postash may be replaced by 3 kg amount of Suphala (15:15:15 NPK). Compost is prepared by soaking the straw chips for 24 hours. Drain out the excess water and keep 2/5th of soaked straw apart. Rest 3/5th of straw is mixed with fertilizers thoroughly.

Pasteurization of substrate

The use of an autoclave to sterilize substrate is not recommended since sterilization also kills beneficial microorganisms, which are present in the substrate. In addition, nutrients in the compost are broken down by sterilization into forms more favourable for the growth and development of competing microorganisms. Thus, substrates that are sterilized are easily contaminated unless spawned under very aseptic conditions, as in media and spawn preparation.

Pasteurization preferably at 100 ^{0}C is more acceptable because the cost is lower and substrates are less susceptible to contamination. The substrate is steamed for 2-3 hours, depending on the volume and the size of the bags. When using a lower temperature (60-70 ^{0}C) as in the case of room or bulk pasteurization, the substrates, whether in bulk or already packed in bags, are steamed for at least 6 to 8 hours.

Spawning

Spawning is carried out aseptically; preferably using the same transfer chamber or the same inoculation room as is used in spawn preparation.

Grain or sawdust spawn is commonly used to inoculate the substrate in bags. With grain spawn, the bottle is shaken to separate the seeds colonized with the white mycelium. After lifting the plug and flaming the mouth of the bottle, a few spawn grains are poured into the substrate bag. Both the plug of the spawn and the plug of the compost bag are replaced and the next bags are then inoculated. The newly inoculated bags are slightly tilted to distribute the grains evenly in the shoulder area of the bag around the neck.

For sawdust spawn, the spawn is broken up with an aseptic needle. A piece of the spawn may then be transferred, using a long flat-spooned needle especially designed to scoop the spawn. One bottle of grain or sawdust spawn in a 500-ml dextrose bottle is sufficient to inoculate 40 to 50 bags.

The highly industrialized method involves bulk-pasteurization and bulk-spawning before the substrates are distributed in beds similar to those used for *Agaricus*. The system is labour saving but requires more complex equipment. Bulk material processing and handling are highly risky for tropical mushroom cultivation due to the risk of contamination.

Incubation

The spawned compost bags are kept in a dark room until the mycelium has fully penetrated to the bottom of the substrate. In 20 to 30 days, depending upon the substrate/substrate combination, the substrate appears white, due to the growth of the mycelium. The bags are kept for an additional week before they are opened to check that the mycelium is mature enough to fruit.

Most strains of the mushroom form primordia after 3 to 4 weeks of mycelial growth. The bags are opened, to initiate fruiting, inside a mushroom house.

Fig. Growth of *Pleurotus* on wheat straw

Cultivation of oyster mushroom may be achieved in following methods

The cultivation of Oyster mushroom or Dhingri mushroom or Pleurotus spp is relatively simple. The agro-climatic conditions prevailing in northern India are conducive for cultivation especially during the months of September to March when the temperature is 20-30 ^{0}C and relative humidity is 70-80%. The production gradually decreases during peak periods of cold from 25 December to 14 January. Even during this period, cultivation of cold tolerant strains can be adopted. There are three different method of cultivation of oyster mushroom.

Rope cot method

Oyster mushroom is generally grown on the nylon netting on rope cots of 2 x 1 metre size. The optimum temperature for growing Oyster Mushroom ranges between 20-30 ^{0}C.

Over the rope cot nylon mosquito net is spread. The straw mixed fertilizers are then spread over the nylon netting. Over this layer of fertilizer mixture, put layer of straw chips (without fertilizers) from approximate 1/10th amount of straw chips. Spread the spawn over this layer and cover the spawn with a layer of same amount of straw chips.

Repeat the first cycle of laying the compost and spawning in same way, described above. Finally cover the spawn with rest of the straw chips.

Nylon netting bag method

A new technique has been innovated, which is quite simple, easy as well as cheap to be adopted for cultivation. In this method housewives who grow vegetables in the kitchen garden can grow Mushroom in the Nylon netting bags. The nylon bags may be hanged anywhere, with the availability of growing condition of oyster mushroom.

The compost for this method may be prepared with formula provided in substrate column, as per the requirement of the compost. Upon 1 kg of straw chips of 1-2.5 cm size, there is a requirement of either 200 g of Suphala (15:15:15 NPK) or 66 g of Urea, 166 g of Superphosphate and 60 g of sulphate of potash.

The nylon netting bags of 45 x 30 cm are filled with the compost material and a layer of unfertilized straw chips is layed as described above. Over this layer spawn is added. The process is repeated for both laying of compost as well as for spawning as stated in the above method.

These nylon-netting bags are then put inside a plastic bag having several holes punched in it, in order to provide proper aeration. The mouth of bags is closed and hanged through the ceiling, where the suitable growing conditions are available.

After an interval of few days open the plastic bag whether it is needed for watering. Nearly after a week pin heads start to appear at this time plastic bag should be removed other operations are some as in the method I. About 500-800 gm of Mushroom can be obtained from a bag.

Rectangular block method

In this method a wooden tray mould without bottom of 50 x 33 x 15 cm size is taken. One square meter piece of polythene is spread so that it forms the bottom of above mould and loose edges hang out from the tray. Wet chopped paddy straw as described in poly bag method is filled making a layer of 5 cm thickness. Now the spawning is done, ensuring a uniform distribution of spawn. In each block approximately 100 g of spawn is necessary. Another layer of 5 cm thickness is layed over this layer and spawning is done in same way as above. Finally third layer is layed over the second and spawning is repeated as above. At this stage spawn is covered with more wetted straw. The straw compressed firmly by hands or wooden board. Loose hanging edges of plastic sheets are folded over the straw blocks and fasten with a string. Now the rectangular wooden board is removed from block. Then the blocks are placed in shelves in a room having relative humidity not less than 80% and temperature 24-26 ^{0}C.

Fruiting

A mushroom house should be used and its size will depend on the number of bags prepared at any one time. The house may be built of thatch or wood or concrete. Air vents on the upper walls should be provided for the ventilation, required for the development of the sporocarps. At the same time a small amount of light should be provided inside the house. The walls may be covered with plastic or foam sheets to increase the relative humidity (80-95%) in the production house.

Shelves, made from bamboo or wood may be used, which should be lined both sides of the house. The shelves should be kept at 40-50 cm spacing between them. They should be strong enough to hold the bags or blocks containing the compost.

The bags are opened by removing the plug and the PVC pipe neck, then rolling down the mouth of the bag. Alternatively, the mouth portion may be cut off with a razor blade, or the bag may be slit either criss-crossed at four to six places, or simply slashed lengthwise. When following the latter technique, the bags may be suspended with a rope or string. When using blocks instead of bags, the blocks are opened either completely or with only the surface or upper portions exposed.

Fruiting requires an appropriate temperature range (20-28 ^{0}C), ventilation, light moisture and humidity (80-95%). To provide moisture, daily watering of the substrate is required. Watering, however, should not be so excessive that the substrate becomes waterlogged.

If the temperature inside the house rises to more than 30 ^{0}C, a light water mist should be frequently used to lower the temperature and hasten fruiting. Doors and windows may also be opened, especially at night, to allow the cool night air to enter.

Approximately 3 to 4 days after opening the bags, mushroom primordia will begin to form. Mature mushrooms should be ready for harvesting in another 2 to 3 days. If the substrate has not yet been completely colonized, the onset of fruiting will be delayed.

Harvesting

To harvest the mushrooms, they should be grasped by the stalk and gently twisted and pulled. A knife should not be used. If kept in a refrigerator or in a cool place, the mushrooms should remain fresh for up to 3 to 6 days.

After harvesting from the top end of the bag, the other end may be opened to allow fruiting. The two ends are sometimes opened and allowed to fruit at the same time. After harvesting from the end portions, slits may be made on the central portion of the bag so that more mushrooms can develop. When a sawdust substrate is used, the harvested surface may be scraped lightly to expose a new surface for fruiting. As long as the substrate appears white, mushrooms will continue to form under adequate environmental conditions. When it appears colorless and soft, it is time to remove the bags from the house.

Yield

Yield ranges from about 100-200% of the dry weight of the substrate and depends on the substrate combination as well as the way in which the substrate has been managed during the fruiting season. From personal observations, the richer the combination and the whiter and denser the mycelium, the greater will be the mushroom yield.

To increase yield, the most common supplement used is urea or orchid fertilizer dissolved in water (100 gm in 100 liters water). Using a plastic mist sprayer, the solution is sprayed on the surface immediately before fruiting.

Storage

The freshly harvested mushrooms have a storage life of only 24 hours at room temperature and as such should be marketed within that period.

CULTIVATION OF *PLEUROTUS SAJOR-CAJU* AT A GLANCE

Materials required:

- Thatched hut/polythene chamber/ mud or pucca house
- Dry paddy straw or other agro waste- 100 kg
- Horse gram powder 4 kg
- Spawn bottles
- Polythene bags 1 kg
- Water sprayer

Method of cultivation

A. Polythene bag method

(i) Take dried paddy straw

(ii) Chop the straw into 1-2 cm bits

(iii) Soak the chopped straw into water for overnight

(iv) Drain off the excess water

(v) Add horse gram powder @ 8 g/kg

(vi) Add spawn @ 30g/kg

(vii) Mix all the constituents

(viii) Fill the mixture into polythene bags; make holes

(ix) Incubate the bags at room temperature between 21-35 ^{0}C for 15 days

(x) Cut open the polythene bags on the sides without distributing the bed

(xi) Spray water over the bags twice a day

(xii) Observe the crop after 22 days of spawning (after 7 days of opening polythene bags)

(xiii) Harvesting is done in three stages leaving a gap of 7 days between each harvesting

B. Rectangular block method (each block require 100 gm of spawn)

(i) Take wooden tray mould without bottom of 50 x 33 x 15 cm size

(ii) One square meter piece of polythene is spread so that it forms the bottom of above mould. And the loose edges hang out from the tray

(iii) Fill wetted chopped paddy straw as described in poly bag method making a layer of 5 cm thick

(iv) Scatter spawn uniformly

(v) Lay another 5 cm thick layer over this and spawn in same way as above

(vi) Finally lay third layer over the second and repeat spawning

(vii) Cover spawn with more wetted straw

(viii) Compress it firmly by hands or a board

(ix) Fold loose hanging edges of plastic sheets over the straw block, fasten with a string. Remove the rectangular block from the block

(x) Place block on shelves in a room having R.H. not less than 80% and temperature 24-26 ^{0}C

(xi) Observe for white cottony growth, which appears after 12 days. At this movement straw becomes compact.

(xii) Now remove the polythene sheet cover

(xiii) Harvest the mushroom, which comes around 18-20 days after spawning. It should be harvested when cap starts becoming folded

(xiv) Consume fresh or dry it in sunshine.

Chapter 10

CULTIVATION OF SHIITAKE MUSHROOM (*Lentinus edodes*)

Besides the button and oyster mushrooms, the shiitake or Japanese forest mushroom (*Lentinus edodes*) is the third most widely produced mushroom in the world, which has increased faster than any other specialty mushroom.

The shiitake mushroom is a large, umbrella-shaped mushroom that is dark brown in colour and is prized both for its culinary and medicinal properties. Proven medicinal benefits include antiviral, antifungal, and anti-tumor effects. It is capable of generating stamina, curing cold, improving the blood circulation and lowering the blood pressure. It has also been shown to stimulate the immune system thereby acting against the cancer cells.

Fig. Fruit body of *Lentinus edodes*

The real shiitake cultivation began in Japan centuries ago when wild shiitake was collected from the forest. Today Japan accounts for approximately 80 percent of the total annual world production of 200,000 metric tonnes. However, it is reported that the beginning of this mushroom was from China, from where it was introduced to Japan by Chinese farmers during 18th century.

Nutrition in shiitake mushroom

Shiitake mushroom contains all eight essential amino acids in better proportions than soy beans, meat, milk, or eggs as well as a good blend of vitamins and minerals including vitamins A, B, B_{12}, C, D and Niacin. In addition, shiitake mushrooms are a popular source of protein becoming a major staple diet in the world population.

When cooked, shiitake imparts a full-bodied aromatic but distinctly pleasant flavor to the dish while maintaining its own original colour and chewy texture. Shiitake is easily dried and rehydrates well.

Shiitake mushroom may be used as a meat substitute in vegetarian dishes and is valued for its full-bodied flavour, dark colour and meaty texture.

CULTIVATION PRACTICES

Shiitake mushroom may be cultivated using wild-simulated or artificial conditions. Wild-simulated mushroom enterprises generally use white oak logs under shaded forest conditions and are best suited for the small-scale producer with an operation of 5,000 or fewer logs.

Typical small-scale wild-simulated shiitake operations produce mushrooms on logs under forest shade conditions. Suitable growing logs are hardwood species and are four to eight inches in diameter and three to four feet long. Tree species desired include the white oak family, elm, sweetgum, and yellow poplar. White oak is often cited as the most desirable species. Tree species are chosen with consideration given to moderate bark thickness as mushroom "pins" must be able to push through the bark at the onset of fruiting. The bark also serves as a barrier to other fungi. Another consideration essential to successful shiitake production is bark retention, which may be accomplished by cutting trees during dormant months (e.g. October through February). Logs should be handled gently to preserve bark and should be inoculated within two weeks after cutting to avoid infection from other fungal species.

A shady area (at least 60 percent shade) should be selected for stacking and storing inoculated logs. This production area, also known as a growing yard, should have enough space to move logs and conduct other production operations.

LOG METHOD OF CULTIVATION

Shiitake or Japanese forest mushroom (*Lentinus edodes*) can be grown on a variety of currently underutilized logs. There are six key cultivation steps, each of which requires careful attention:

Obtaining viable spawn (inoculum) in pure culture and storing it until use.

Shiitake mushrooms are grown from spawn. Mushroom spawn may either be purchased or prepared as suitable for the area. Different strains are better suited for different environmental conditions.

The pure culture spawn is prepared by mixing rice bran with sawdust. The mixture of rice bran and sawdust is well autoclaved in order to sterilize it. After autoclaving, the mixture is inoculated with mycelium of the species and kept at the temperature between 24-28 ^{0}C

Spawn may also be prepared using wedged round rod shaped of wood of various species, particularly oak. These pieces are generally chosen having about 1.5 cm in diameter and 2.0 cm in length. The wood pieces are autoclaved and inoculated as above.

Obtaining and preparing logs for cultivation

The species of tree selected for shiitake cultivation is important. Trees are felled during the dormant season and cut to length (3 to 5 ft.) prior to inoculation. Cut should be made at the right time of season, as if wrong time chosen, the bark strip off easily and chances of contamination of other wood fungi are greater. In such case shiitake mycelium has to face more competition from inhabitants, as there is no such procedure for sterilization of logs.

Inoculation

Inoculation is the introduction of the live spawn into the log. It is perhaps the most time-consuming operation in shiitake production. The process of inoculation begins by taking harvested, cut to length logs and drilling rows of holes about 6 inches apart along the length of the log. Rows should be offset and spaced about 2 inches apart to allow maximum room for mushroom growth. Logs average 35-40 holes each. Hole depth and diameter will vary depending on whether dowel plug or sawdust spawn is used. Dowel spawn generally requires a 5/16-inch diameter hole drilled to a depth of 1-1½ inches. Sawdust spawn will require a slightly larger diameter hole and less depth. Dowel plug spawn is driven into each hole using a hammer; sawdust spawn requires a special inoculating tool.

To prevent bacterial or fungal competitors from entering the log, each inoculation hole should be sealed with a thin coat of hot wax using a sponge. Cheese wax works well and remains pliable over a wide temperature range. Logs are then labeled with the inoculation date and spawn strain using a metal tag hammered into the end of the log.

Laying the logs

Logs are often laid to favour fungal growth side-by-side, propped up at a slight angle in a well-drained, shaded area with single logs placed crosswise between rows. Logs may be turned reversing the ends every 2 to 4 months to encourage uniform water distribution and fungal growth.

Management of logs

Shiitake is capable of fruiting only after the fungus has completely colonized the log, which generally completes in 6 months to

Fig. Laying of logs

18 months. Fruiting occurs primarily in the wet, cool seasons (spring and autumn) but adequate rainfall during the summer months can stimulate fruiting, especially with certain strains or varieties. Fruiting occurs usually at 12-20 ^{0}C temperature. When conditions are favorable for fruiting, logs can be arranged in rows in an upright position on well-drained, shaded ground. An upright stacking arrangement facilitates harvesting the mushrooms.

After inoculation, logs should be stacked in the shade or covered with 60 percent shade cloth. Logs may either be stacked log cabin or lean-to style. Proper moisture content is critical for proper incubation and should be monitored regularly. Logs should never dry out, but should not be so wet as to produce mold; it is important to allow the bark to dry out between watering. Good air circulation will help to prevent molding. During the incubation period, water logs for 12 hours once or twice a month. Logs may be watered either using a sprinkler or by immersion in a livestock feed trough. Expect to water logs more frequently during hot, dry summer months.

Logs will begin to fruit between 6 to 18 months after inoculation and will continue to produce mushrooms for about three to five years depending on log diameter. Growers report the second and third years after inoculation as the most productive, double the production of that during the first and fourth years.

Fruiting is near when rings of white mycelia appear on the end of logs. At this point, logs may be stacked lean-to style to improve ease of harvesting. While mushrooms may be forced to fruit by soaking logs, better yields will result if the spawn is left to incubate slowly, allowing spring and fall rains to soak the logs. This strategy will also minimize your management and produce good spring and fall harvests, but very little production during summer and winter months. Year round production may be achieved by simulating fall and spring conditions throughout the year. However, this technique requires much more labor and management and requires the use of indoor climate controlled growing rooms. Mushrooms should be harvested when their caps are about two-thirds open, cutting the stem flush with the bark with a sharp knife. Mushrooms may be stored in a cardboard box for up to a month at 36-41°F. Mushrooms may also be dried whole or sliced before marketing but you will receive substantially lower prices than fresh.

Fig. Lentinus mushroom growing on log

Harvesting the crop

Once mushroom formation has begun, shiitake often matures in 2 to 7 days. The mushrooms can be either snapped off cleanly or cut with a knife at the log surface. Fresh mushrooms intended for market should be refrigerated immediately after harvest. 100 kg natural log can produce 10-15 kg fresh mushrooms over this period.

POLY BAG OR HIGH INTENSITY CULTIVATION METHOD

While many wood utilising mushrooms have traditionally been cultivated on hardwood logs outdoors in the natural environment, an alternative more intensive and regulated cultivation technique has been developed in several laboratories in the recent past. This high-intensity cultivation employs compressed sawdust /grain blend logs or blocks for growing medium and indoor growing environments.

This method also requires a much larger capital investment including dedicated growing houses with climate control, sawdust logs, and special equipments. Because of the large initial capital investments required, high-intensity mushroom cultivation incurs much greater economic risk than in wild-simulated operations.

Small-scale shiitake operations are often a family affair and the time spent together is frequently cited as a motivating factor to begin and continue production. However, even the motivation of family time together will not keep the enterprise in production long if it does not realize at least a modest profit.

In this innovative approach various hardwood sawdust or wood chips supplemented with nitrogen-rich additives such as rice bran are mixed together and then compacted into special autoclavable polypropylene bags of various dimensions. The bags are autoclaved for sterilization, then allowed to cool to 20 ^{0}C and inoculated aseptically with the desired amount of spawn. This stage demands complete sterile handling techniques and any relaxation of standards allows microbial contamination with concomitant financial losses. The inoculated bags, sometimes known as space bags or artificial logs, can then be placed to growing rooms with controlled environments giving accurate humidity and temperature conditions.

Following inoculation the bags are stacked on trays or suspended from wires for several weeks for mycelium run through the sawdust mix. Mycelium while run secretes several enzymes, which degrade the complex macromolecules (lignin, cellulose, hemicellulose) of the substrate. The breakdown products being absorbed by the advancing mycelium. When the mycelium has reached maturity the log is given a cold temperature shock for 12-24 hours, restacked and the bag opened and within a few days the mushrooms develop.

Overall, this new method greatly shortens the production time and gives much higher yields. Using natural log cultivation the time from spawning to harvesting of mushrooms can be between eight months to one year with complete exhaustion of the log up to three years.

100 kg sawdust plus supplement can produce 80 kg fresh weight of mushrooms. This represents at least an 8-10 fold greater production than the natural log method.

STORAGE AND MARKETING

Main considerations for mushroom buyers include freshness, colour, and shape. Signs of age include withering and darkness from oxidization. Shiitake mushrooms can generally be refrigerated for 4 to 5 days with little product deterioration. Local buyers and outlets for the small-scale producer include restaurants, bed and breakfasts, vacation resorts, organic retailers and markets, supermarkets, and farmers markets.

Buyers value consistent supply and generally place a high value on their relationships with producers. This may translate into a small premium for the reliable small-scale producers. If small producers are unable to find local buyers, wholesale buyers (who may also be larger producers) will buy dry product, but offer about half the price per unit of fresh mushrooms.

The smaller, wild-simulated producer may have a quality advantage over high-intensity cultivated mushrooms. Some buyers report that mushrooms grown on sawdust blocks are lower in quality with regard to taste, shelf life and increased cooking shrinkage. Given quality advantages, without the insurance of quality year-round delivery, buyers will be more likely to purchase from large-scale industrial growers. Grower cooperatives and year-round producers can greatly aid in the process of helping small producers to deliver reliable, fresh supplies of mushrooms to market.

Considerable profits may be realized with the development and effective marketing of value-added shiitake products including sauces, soups, as well as fresh and dehydrated shiitake products.

RISKS ASSOCIATED WITH SHIITAKE CULTIVATION

Producers using forest-grown cultivation techniques report many risk factors associated with growing shiitake mushrooms. These include adverse weather conditions, pest damage, securing a consistent source for viable spawn, and marketing challenges. Moisture during the summer growing months is critical for a quality product and cooler, damper summers are favored. Extended hot, dry periods during summer months will negatively impact mushroom quality and yield and may significantly reduce price commanded at market. During extended dry periods, humidity can be controlled under forest conditions through use of irrigation, but at significant cost and assuming a clean, reliable water source in proximity to the production site.

Pests of shiitake include termites, slugs, squirrels, mice, birds, and deer. In some areas of the region, black bear may visit the production site inflicting considerable damage to logs and equipment in a short period of time.

Economical consideration of shiitake cultivation

Before starting any new enterprise, it is important to gather and analyze current cost information. In addition, some initial study of potential markets should be conducted; thereby assuring that your investment will be warranted.

Chapter 11

CULTIVATION OF PADDY STRAW MUSHROOM (*Volvariella* spp.)

Since the "paddy straw" is mostly used as the substrate for cultivation of *Volvariella* spp., therefore all species under the genus *Volvariella* are generally called "paddy straw" or 'straw' mushrooms. San Antonio and Fordyce (1972) confirmed, that straw mushroom derives its name from the substrate on which it originally was grown. However, these may be found growing on other substrate in the natural condition.

Fig. Fruit body of *Volvariella volvacea*

World scenario of straw mushroom production

Straw mushroom accounts for approximately 3% of the total world production of edible mushrooms. It is well suited for cultivation in the tropics because of its requirement for higher production temperatures. It is preferred as a kitchen garden crop because it is very delicious and nutritious. It is, however, less attractive commercially owing to very low yield potential against per unit weight of the substrate and an extremely short shelf life.

Nutritive value of paddy straw mushroom

Paddy straw mushroom contains a high quality protein with important amino acids including lysine. Vitamins and minerals are also present in quite reasonable amount. Paddy straw mushroom consists of moisture (90.4 %), fat (0.25%), protein (3.9%), crude fibre (1.87%), ash (1.1%), and phosphorus (0.1%), potassium (0.32%). *Volvariella* species also contains iron, calcium, thiamine, riboflavin, niacin and ascorbic acid @ 1.7, 5.6, 0.14, 0.61, 2.4, and 18 mg/100 g fresh weight, respectively.

Species under Genus Volvariella

The genus *Volvariella* consists of several species including *V. bombycina, V. diplacia, V. esculenta, V. speciosa, V. terastius, V. volvacea*. The species *V. bomycina* is distributed in near the **Lucknow** (Uttar Pradesh) growing on decayed logs. They are large in size, white and centrally stipitate. Pileus goes around 8.0-20.0 cm in diameter. Stipe is usually 5.0-15.0 cm long, 1.0-1.5 cm thick tapering slightly at the tope, base enlarged, usually white but tan towards the base having large volva. *V. speciosa* is also distributed near Lucknow.

V. diplacea is found distributed in Baroda, West Bengal and other parts of India. It is found growing on the rotten paddy straw having a unique character of absence of pigmentation. It has short diameter (8.0 cm) of pileus of hemispherical in shape. The stipe is comparatively long and narrowed upward (2.5 cm broad at the base and 0.8 cm at the top). *V. terastius* is distributed in West Bengal and found growing on the rotten paddy straw heaps.

V. esculenta is found in southern states including Tamil Nadu, found growing on the decaying logs. Pileus is 14.0 cm in diameter. Stipe is medium in size ranging from 8.0 to 12.0 cm. The pileus is white, stipe is creamish in colour.

The *V. volvacea* which is most likely and cultivated species among the paddy straw mushroom, is found growing on rotten paddy straw heaps in the natural conditions. It is distributed throughout India. The pileus is 5.0-12.3 cm in diameter, having a distinct radial sepia coloured streaks up to the middle of the pileus. The pileus is soft in touch and found sometimes with split margins. The stipe is central, cylindrical, attenuated upward, 8.0-14.0 cm long, whitish in colour, ending below with a solid bulbous base. The volva is well developed and persistent.

Historical Background

Historically, paddy straw mushroom was grown in china during eighteenth century; therefore it is called Chinese mushroom and Tributary mushroom. The Nauhua mushroom so called as it first grown in the Nauhua temple of China by the Buddhist monks. In India the scientific culture of Paddy straw mushroom was started at Compbatore.

Cultivation of Volvariella is believed to be started in China as early as in 1822 (Chang 1977). In 1930's, straw mushroom cultivation began in the Philippines, Malaysia, and other Southeast Asian countries (Chang 1982).

Requirements and substrate for growth

In the coastal region the continuous heavy rains decomposed the paddy straw in the Farmyard. *V. volvacea* is found growing on such decomposed straw. The decomposition in coastal areas takes place during month of June to October when the temperature ranges between 30-35 ^{0}C. However, in the plains it is cultivated in the months of March to October, as these months experience such high temperature.

Though paddy straw mushroom is most suited for little high temperature, but it may be grown in winter season also. In the winter season temperature of bed may be increased by the addition of leaf materials from Tamarind and Neem leaves, either dried or fresh. The ratio of neem or tamarind leaves is appropriate, if it is kept at 1:10, e.g. one part of neem/tamarind leaves and ten parts of paddy straw.

Paddy straw mushroom being the popular in the country people, it is now days grown in other than paddy straw on cotton waste and water hyacinth waste. Banana leaves are also used for the cultivation of paddy straw mushroom.

VARIOUS METHODS OF CULTIVATION

There are various methods through which straw mushroom cultivation is performed, in order to fulfill the ever-increasing requirements:

Rectangular block Method

Chop the straw either manually or mechanically into bits of 3-5 cm in length and pack in gunny bags. Boil the water in a drum. When the water comes to boiling, place the gunny bag along with the straw in the boiling water and boil as such for 15 to 20 minutes. Then remove the gunny bag from the drum and leave as such for 8 to 10 hrs to drain the excess water and also allow the straw to cool. Care should be taken that the bag is not open till the time of block making, as this will contaminate the boiled straw. The desirable moisture content of the straw can be tested by squeezing the straw in between the palms, and see that droplets of water do not trickle out from the straw.

Another method of pasteurisation of the straw is by steaming. This method requires little modification of the drum. Punch a small hole in the lid of the drum, and while boiling the straw; seal the surrounding of the lid with a rubber tube. The chopped straw should be previously wetted and excess water drained off. Place a few stones in the drum and pour water only at the level of the stones. Boiled the wetted straw by keeping it in a bamboo basket and place the basket over the stones inside the drum. Close the lid of the drum and seal the rim of the lid by means of a rubber tube. The steam generated from the boiled water will pass through the straw and pasteurise it. After boiling, transfer the straw into a previously sterilized gunny bag and leave it as such for 8-10 hrs to let it cool.

Take a wooden frame and place on a smooth floor. Place jute ropes, two vertically and one horizontally. Line the frame with a plastic sheet, previously sterilized by dipping in boiled water.

Fill approximately 5 cm of boiled straw and compress it with the help of a wooden lid and sprinkle spawn over the whole surface.

After the first layer of spawning, put another 5 cm of straw and again sprinkle spawn over the surface, compress it as in the first layer. In this way, continue to sprinkle spawn over the layer of straw for 4 to 6 layers till the straw is in level with the top of the frame. Only (1) one packet of spawn should be used for one cube or block.

The plastic sheet is now folded over the top of the frame and tied down with help of jute ropes previously placed below the plastic.

After tying, the frame can be removed and what is left behind is a rectangular block of straw. Punch holes (2 mm diameter) on all sides of the block for aeration.

Place the block in an incubation room, side by side in shelves in single layer only taking care that they are not place directly on the floor or on the top of each other as this will generate excess heat.

The temperature of the block should be maintained at 25^{0}C. This can be noted by inserting a thermometer into the holes of the block. If the temperature rises above 25^{0}C, it is advisable to aerate the room, and if the temperature falls, the room should be slowly heated up.

It takes 12 to15 days for the spawn to spread throughout the straw and when the entire block has been completely white, it is a sign that spawn running is over.

After spawn running, remove the ropes as well as the plastic sheet from the block. Tie the block vertically with coconut rope and hang it in a cropping room. From this stage onwards, the relative humidity of the room should not be less than 85%. This can be maintained by periodically spraying water on the walls and floor of the room. If it is a cemented floor, it is advisable to pour water on the floor so that water always remains on the floor. If the block shows signs of drying light, spraying can be done with help of sprayer.

Within a week to 10 days, tiny pinheads will be seen on the surface of the block and these will grow into full-size mushrooms within a day or two.

When fruits bodies start forming, the requirement of air is increased. Therefore, once fruit bodies start forming, it is essential that there is an exchange of fresh air every 6 to 12 hrs, by opening the ventilator provided at the front and back of the room.

The fruits bodies (mushrooms) are ready for picking just when the periphery of the caps starts turning upwards. This will be evident as small crinkles appear on the side of the pileus (cap). To harvest the mushrooms, take hold of the stipe (stalk) at the base with thumb and forefinger and with a gentle anti-clockwise twist, detach the mushroom from the straw without disturbing the straw or any small mushroom growing alongside. Do not use knife or scissors for harvesting. The block will again come to fruiting after about a week.

Heap Method

The cultivation of paddy straw mushroom may be completed in following steps:

1. Management of Paddy Straw (the substrate)

The leafy top of the paddy straw bundles is removed. The bundles are soaked in water for 18-24 hours, in a big container. The bundles are put in such a way that they are completely immersed in water. In order to facilitate the full emersion of bundles, weight of bricks blocks, stones, etc given over them. After the said period, the straw is taken out from the pool of water and kept in the hanging condition, at least for one hour to ensure proper draining of water. Then straw is stacked on a bamboo frame, following the method described below.

2. Heaping the substrate

The bundles are staked in a layer on the bamboo frame, which is held at 30 cm above the floor by a number of supports of bricks. Now a days the farmers use rope cots for the purpose. In order to prevent rotting of the ropes, nylon ropes are preferred instead of hemp ropes.

The straw stacks are made in such a way that the butt ends of the bundle remain one spreading lengthwise close to each other. The number of bundles should be equal to the length of the straw, to ensure a square bed. Above this first layer of straw, the second layer of straw is made by placing the bundles having the butt ends on the opposite side.

3. Spawning the substrate at the second layer

At the end of these two layers, the small bits of spawn are scattered towards the border of the stack. In order to give a quick start of the spawn, a small quantity of *Besan* is sprinkled over the spawn bits.

4. Making third layer of heap

After the spawning at second layer, a third layer is made. The third of bundles of bundles is placed at the right angle with the *criss-cross* fashion to the above layer. The fourth layer is arranged on the third layer with the opposite butt end fashion (same as second layer).

5. Spawning the substrate at the fourth layer

Same as the above, spawning is done with small bits of spawn scattering towards the margin of stack. The small quantity of *Besan* is also sprinkled over the spawn bits.

6. Stacking and spawning of fifth to sixth layer

Same as the above stacking is done for fifth and sixth layer. The spawning is done in same way described above. At the end of spawning *besan* is sprinkled again in order to ensure proper growth and development of the spawn.

After the sixth layer of spawning spawn is covered with a layer of wet straw. Then the straw pressed in order to make compact. For six layer bed, about 350-400 gm of grain spawn is required. In order to increase the size of fruit bodies, some chemical fertilizers like Ammonium sulphate @ 560 gm and SSP @ 112 gm should be applied at after every spawning in the equal parts of amount. Finally the bed is covered with polyethylene sheets. After seven-ten days of spawning pinheads appear to start. At this the polythene cover should be removed. The pinheads take next few days to grow in full size of Mushroom. Throughout the growing time, the temperature should be maintained at 30-35 ^{0}C. It should be ensured that temperature should not go lower than 30 ^{0}C.

Fig. Growth of fruiting body

The best time of picking such fruit body is when the volva is about the rupture. It should be harvested in three different stages by gentle twisting of the fruiting bodies.

Hollow bed method

This method of cultivation was described by Krishnamohan and Jaya Rajan in 1978. In this method, a cement pot of 30 cm diameter is used. Around this pot, the straw twists arranged in such way make a bed. After the completion of bed, it is pressed to make it compact. Now the cemented pot is removed leaving a hollow centred bed. The spawning is done in both the sides of inner and outer side. This method is useful as it provides more spawning and growth for the mushroom.

Poly bag Method

This method was described by Bahl in 1982, include soaking of straw chips in water for 24 hours; simultaneously soaking of waste newspapers of $1/4^{th}$ of the straw size for same period. After soaking, soaked straw and pulp of paper are mixed together and filled in polyethylene bags. Small holes are made in the bag in order to ensure proper aeration and removal of water from the bag. The fully packed and punctured bags are put in the growth place having temperature between 35-40 ^{0}C. After a week of development, the polythene cover is removed gently so that the lump of the mixture is not broken. Now this lump is placed on a wooden plank and water is sprinkled to ensure a proper humidity of 80-90%. The pinheads appear after 10-15 days of spawning.

Alicbusan's method

This method was described by Alicbusan in 1979, is suitable for the village people. In this method the straw is soaked for three hours and and then it heaped on a bamboo frame over polythene sheets. Simultaneously the crumpled newspaper sheets are soaked in a container having 3 g urea/ 4.5 litre of water. The soaked paper is planted along with the spawn. The mushroom spawn and soaked paper are distributed on the top of layer in thumb-sized pinchs. Spawning should be done in a way, that margins of 5 cm are left from each side. The spawn in buried with 4 cm deep layer of newspaper sheets and immediately after spawning bed should be protected by a transparent plastic sheet. Watering is done on each bed frequently according to the humidity in the growth chamber. The pinheads start to appear after 7-10 days of spawning.

Cultivation in Cotton Waste

In addition to straw, many agricultural by-products and waste materials have been used to produce the straw mushroom. These include paddy straw, water hyacinth, oil palm bunch, oil palm pericarp waste, banana leaves and sawdust, cotton waste and sugarcane waste.

In recent years, cotton wastes (discarded after sorting in textile mills) have become popular as substrates for straw mushroom production. Cotton wastes give higher and more stable biological efficiencies (30 to 45%), earlier fructification (four days after spawning) and harvesting (first nine days after spawning) than that obtained using straw as a substratum. Semi-industrialization of paddy straw cultivation on cotton wastes has occurred in several countries including Hong Kong, Taiwan and Indonesia.

Pest and disease control

If the bed is infected with fungus the production will be reduced. Common mold are *Aspergillus* spp., *Rhizopus stonelifera rolfsii*. Besides *Coprinus artatus* and *Sathyrella* spp. are found to grow in special cases. Moulds could be controlled by spraying fungicides like Captan and Zineb. Malathion (@ 0.025%) and Benlate (@ 0.025%) are mixed together and sprayed on the beds in order to destroy pest (mites) and diseases.

CULTIVATION OF *VOLVARIELLA VOLVACEA* AT A GLANCE

Materials required:

- Paddy straw
- Small water tank
- Bricks
- Bamboo frame
- *Besan*
- Spawn @ 350-400 g/bed
- Thermometer
- Polythene sheets

Method of cultivation

(i) Collect paddy straw which is not too leafy, not more than one year old; protect from rain in rainy season; make bundles

(ii) Soak the bundles in water for 18-24 hours in small tank, keeping in view that the bundles are completely immersed in water

(iii) Take out bundles from the tank and drain off excess amount of water

Making of beds

(iv) Make 15-20 cm high foundation of bricks. The size of foundation should be slightly larger than that of bed.

(v) Make wooden bamboo frame of the size of the bed foundation and place it on the top of the raised bed foundation

(vi) Make square beds of one cubic meter size from soaked straw bundles with their butt ends on one side placed length-wise close to each other on the bamboo frame. Make second layer over the first by placing the bundles having butt ends at opposite side, which leads to complete one layer.

(vii) Place small bits of spawn 8-10cm inside the margin, leaving a space of 5-6 cm from each other

(viii) Make the third layer in same way following cress-cross fashion. Place the fourth layer on the third with opposite butt ends

(ix) Place spawn at this layer following the step *VII*

(x) Place another layer of bundles in same way as above

(xi) Spawn this layer too

(xii) Cover the inoculated layer with loose straw

(xiii) Press down the bed

(xiv) Cover the bed with polythene sheets

(xv) Observe for temperature which should run throughout the period between 30 and 35 ^{0}C

(xvi) Remove polythene sheet after 7-10 days of spawning; small buttons appear

(xvii) Harvest the mushroom by gentle twisting of the fruiting bodies

Either consume fresh or dry the mushrooms in the Sunlight or at the temperature of 50-60 ^{0}C and then place in aluminum foil.

Chapter 12
POST - HARVEST TECHNOLOGY

Since mushrooms are perishable and delicate in nature, these cannot be kept afresh for more than 24 hrs. After the maturing of the fruiting body, the deterioration starts with the formation of brown colouration resulting into quality deterioration and loss of marketability. To overcome this problem, especially during peak season, suitable post harvest management/practices are necessary to increase the shelf life and marketability of mushrooms. Initial steps are, proper harvesting time and harvesting stage. Mushrooms are generally harvested after 3 weeks of casing. Button mushrooms are harvested when the cap size is 30-45 mm in diameter, whereas Oyster mushroom is harvested when the fruiting body becomes curled under edges and well-formed gills.

For the purpose of processing, fresh, smooth in appearance, free from major blemishes, dry surface mushrooms are selected. It is worth to mention that closed veil mushroom indicate a delicate flavour, whereas open veil mushroom has a richer flavour. After selection gently wipe the mushrooms with a damp cloth or soft brush to remove occasional peat moss particles attached to fruiting body.

When mushrooms are purchased loose, they must be stored in a paper bags. If purchased packaged, do not open until ready to use; store unused in a paper bags. Storing in airtight containers or plastic bags will cause condensation and speedy spoilage.

Mushrooms need not to be peeled. They should be washed briefly under cold water and allowed a few minutes to air dry. Never soak mushrooms in water to clean them, as they are porous and absorb water. Some mushrooms spoil quickly while others have a longer life span. Shiitake mushrooms can be kept up to two weeks in the refrigerators. Some mushrooms including white button mushrooms, oyster, can be eaten raw. They can be chopped, sliced, quartered, minced, or pureed.

It is important to conserve the right amount of moisture before storing mushrooms. They must not be left uncovered, as they dry quickly. It is also advised that mushroom should not be kept enclosed in moisture-proof wrappings, as they become soggy and decay shortly. A good compromise could be to place mushrooms purchased in bulk in a loosely closed paper bag or in a shallow glass dish covered with a kitchen towel or a lightly moistened paper towel. Leave prepackaged mushrooms in their unopened package. Do not wash or trim mushrooms before storing them.

It is advisable to keep mushrooms on the refrigerator shelf, not in the refrigerator crisper, which tends to be humid for a few days. Unopened, prepackaged mushrooms will stay for up to a week. Whereas those begin to darken (and their caps open), they suggest that they are running with age, however, they could still be used for cooking and flavoring foods.

Dried mushrooms can be kept almost indefinitely if wrapped in plastic or placed in a tightly closed jar and stored in the refrigerator or freezer. They can also be stored in a cool, dark place for up to six months.

Grading of mushrooms

Grading of mushrooms is important for marketing. It is done by segregation of mushrooms into various grade standards as per the market demands. Button mushrooms are graded into Grades A, B and C. The Directorate of Marketing and Inspection (D.M.I.) had formulated the grade standards for dried edible mushrooms as Mushrooms Grading and Marking Rules, 1972 under the Agricultural Produce (Grading and Marking) Act, 1937.

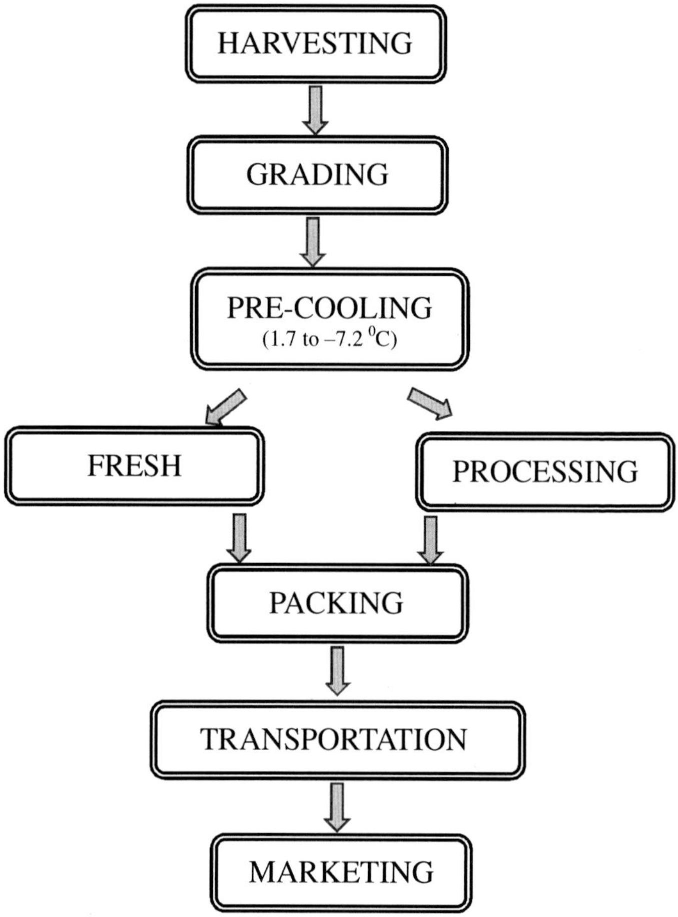

Fig. Post Harvest Practices at a Glance

Pre-cooling and packaging

Other step is pre-cooling wherein the produce is kept in a plastic bag and stored in cooling unit. Vacuum cooling is another cooling system where water existing in cell walls and inter hyphal spaces of produce is evaporated under pressure which lowers the temperature. But it is cost oriented system and involves inevitable loss of fresh weight.

Packing is essential to protect the mushroom during marketing. It is generally packed in polythene bags of 250-400 g for local markets. For long distance transport, pulp board punnets wrapped with PVC films should be used instead of polythene bags. Sometimes, pre-cooled mushrooms are packed in insulated containers having ice in it, so that Mushrooms remain fresh, healthy during long transport.

The effect of pre cooling is lost partially in post pre-cooling period. Hence, for long distant transport, speedy refrigerated transport system is essential in which cartons containing polythene pouches of mushrooms may be transported by refrigerated trucks. For short distance markets, pre-cooled fresh mushrooms are dispatched in wooden boxes with sufficient crushed ice in polypacks. For local retailer's markets mushrooms are packed in polypouches and dispatched by cycle, cycle-rickshaws, auto-rickshaws etc.

In this country mushrooms for the fresh market are generally packed in unlabelled polyethylene or polypropylene hand sealed and manually perforated polybags (20 cm x 15 cm) of 200 g each. Mostly this pouch is made up of 100 gauze polypropylene.

PRESERVATION OF MUSHROOMS

Mushrooms like other fruits and vegetable are highly perishable in nature and need a quick and effective preservation technique. There are some methods suggested below for preservation of mushrooms.

Short-term storage

Fresh mushrooms can be preserved for 3 days if they are kept at or below 4 ^{0}C temperatures in plastic packets of 100 gauze sizes. For the same after the harvesting, mushrooms are washed in a solution of 0.05% potassium metabisulphite for 30 minutes, and then packaged into packets of mushroom weighing 200-400 g and finally stored at 4^0C. The shelf life can be extended from 3 days to one month if the packets are kept in 10-12% brine solution.

Long-term storage

There are several processing techniques such as Canning, Individual Quick Freezing (I.Q.F.), Vaccum Freeze Drying (VFD), Drying, Vaccum Drying, Pickling, Preservation by Steeping and Radiation used for long-term preservation of mushrooms. These techniques are used on the basis of their merits *per se* market demand and end use.

Canning

Canning is an established process of preserving mushroom pieces in brine, butter, oil, vinegar etc. It involves six basic operations like cleaning, blanching, can filling, sterilization, cooling and labeling. Through all these operations, mushrooms are graded, cleaned, blanched (pre-cooked), filled in brine solution of cans and ultimately sterilized with heat and cooled through water spray and labeled for storage. Canned mushrooms form major share of world trade.

In canning mushrooms are washed; stalk is cut close to button; blanched in boiling water having 0.2% citric acid for 3-5 minutes. At this stage there is a loss of 25-30% weight. The cans are filled @ 200g/1 lb can. Hot 2% brine + 2% sugar + 0.3% citric acid solution is added to it. The cans are dehydrated at 80 ^{0}C. Finally the steam is given at 15 lb pressure per square inch for 15-20 minutes.

Individual quick freezing

Individual quick-freezing is another popular processing method followed in large industrial units. In this process, raw materials are washed at processing units after receipt from farms, and then the mushrooms are inspected, sliced and graded according to quality. After that, blanched and water cooled mushrooms are subjected to tunnel freezing stage. At this stage, these are cooled in a system having temp – 40 ^{0}C and the core areas of mushroom pieces acquire a temp around –18 ^{0}C. Subsequently, packed in multi-layer polybags and stored in a cold storage having temperature –20 to –25 ^{0}C.

Button Mushroom with Small Stalk

↓

Wash Thoroughly

↓

Cut the Stalk close to button

↓

Cut the Stalk close to button

↓

Blanch in Boiling water having 0.2% citric acid for 3-5 minutes

↓

Observe weight loss up to 25-30 percent

↓

Fill hot in cans @ 200 g/1 lb can and add 2% salt, 2% sugar, 0.3% citric acid solution

↓

Exhaust at 80 ^{0}C

↓

Steam at 15 PSI for 15-20 minutes

Fig. Flow Chart for Canning Button Mushroom

Vacuum freeze drying

Vacuum freeze drying (V.F.D.) is a further development in mushroom processing technology. In this process the original shape, quality, colour size, texture, freshness properties of thermal sensitive produce are retained. This processing technique involves the cooling of mushroom much below the freezing point i.e. –40 ^{0}C where moisture present in mushroom is converted to tiny ice crystals which further directly sublime into vapour when subjected to vacuum with a slight rise in temperature resulting a dried end product.

In this method, mushrooms are washed properly and dip in 2% sodium chloride solution. Blanching is done in a solution of citric acid @ 0.2% for 3 minutes. Freezing is done at –40 ^{0}C for 30 minutes in a freezer. Then the mushrooms are dried at low temperature and partial vacuum. Packing of mushrooms is done in airtight containers having inert N_2 gas. The product is stored at low temperature.

Drying

Drying is the age-old practice of preserving mushrooms at ambient temperatures. With the advancement of technology, different kinds of dehydration processes have been developed e.g. Sun drying, mechanical drying, air-drying, microwave oven drying, etc. Among these the microwave oven drying is the best method.

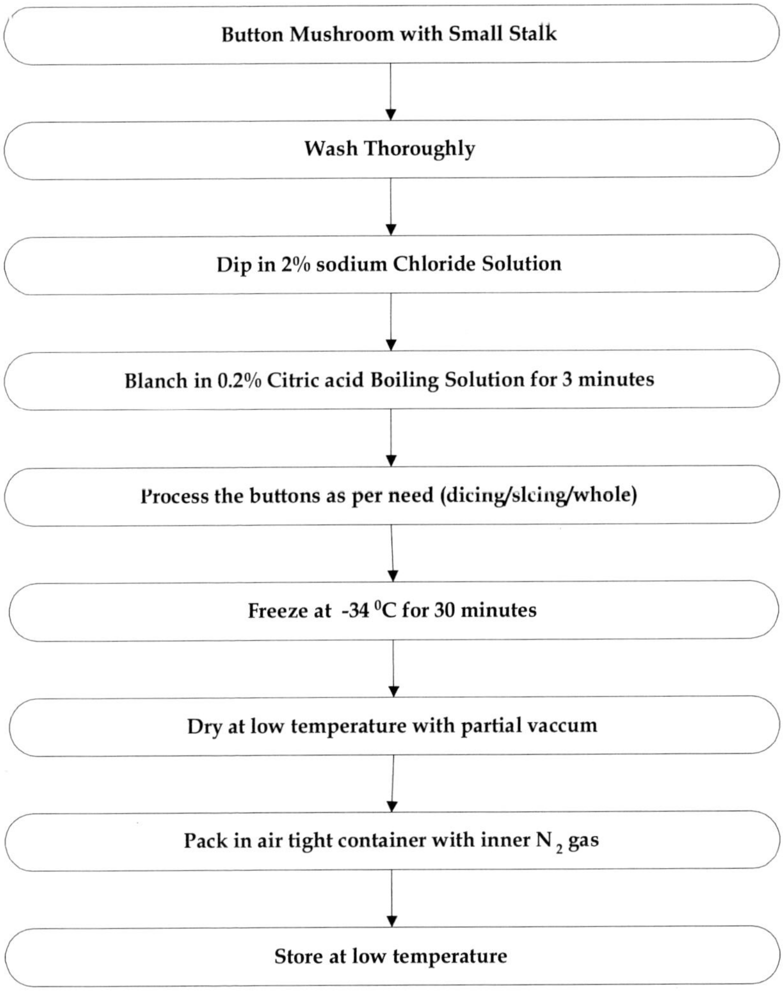

Fig. Flow chart for vaccum freeze drying

Wash Thoroughly

↓

Blanch in 0.2 % Citric acid Solution for 3 minutes

↓

Dip in 600 ppm soulution of Metabisulphite for 5 minutes

↓

Dip in 800 ppm soulution of Sodium chloride for 5 minutes

↓

Dry in Oven at 50-60 ^{0}C to lose moisture to 5%

↓

Seal in cans

↓

Store at low temperature (3-5 ^{0}C)

Fig. Flow chart for Dehydration

Vacuum drying

Vacuum drying process is followed in reduced pressure. In this process, the produce is subjected to a vacuum drier in which steam is present at about 1.0 to 1.5 bar. A vacuum pump is used to reduce pressure inside the product. The end product is obtained after completion of vacuum drying process.

Preservation by pickling

Pickling of mushroom is also a popular method of preserving. It is a more economically viable way during the surplus periods. By this process, mushrooms and different spices/condiments according to preference are fried separately. After that these are allowed to cool and filled in bottles/containers.

Preservation by radiation

Gamma irradiation of fresh mushroom has been found to extend shelf-life, less browning and delay in cap opening. Irradiation with Gamma rays at the dose of 100-150 Krad is subjected to stop the post harvest growth and discolouration/deterioration in the mushrooms.

Preservation by steeping

Different cocktails of chemicals preservatives mainly sodium chloride, citric acid, sugar, ascorbic acid, sodium bicarbonate, potassium metabisulphite, etc have been suggested for steeping preservation. Detailed method is described below:

Wash Thoroughly

↓

Blanch in 0.1 to 0.2 % Citric acid Solution for 3-5 minutes

↓

Cooling

↓

Steeping in 15% salt + 0.5% Citric acid + 200 ppm Potassium Metabisulphite Solution

↓

Fill in Glass Jars

Fig. Flow chart for preservation by steeping

Chapter 13

MANAGEMENT OF SPENT MUSHROOM SUBSTRATE

Natural plant wastes are the basic substrates used for the production almost all kinds of edible mushrooms. Spent mushroom substrate (SMS) is the used substrate after the cultivation of mushrooms. The management of SMS is necessary after the crop harvest, as it may cause various environmental problems including ground water contamination and nuisance. The term SMS is replaced now with post mushroom substrate (PMS), as it is believed that spent word does not fit for this used substrate. The term spent conveys the meaning that it cannot be used further. But recent discoveries illustrates that substrates after one cycle cultivation mushroom can be used for other purposes, including addition in the soil.

Characteristics of Spent Mushroom Substrate

SMS normally contains 1.9:0.4:2.4 percent NPK before weathering and 1.9:0.6:1.0 percent after 8-16 months of weathering. During weathering potassium leaches in significant amount. SMS is away from the category of hazardous waste, as it does not contain a negligible amount of heavy metals. The SMS obtained from various sources usually have conductance in the range of 1.9-8.3 mmhos/cm^3. The chlorine content in fresh SMS varies from 1.5-7.5 kg/tonne, whereas it is 294 ppm for well rotten SMS. The lower chlorine contents are attributed to less chicken manure in the formulation used for cultivation. The pH of freshly collected SMS is 7.28 which increase during weathering. The freshly collected SMS is reported to have density of 0.198 g/cm^2. SMS contains 45% water.

Utilization of SMS for different purposes

Since the SMS has many positive attributes, hence it can be used for various purposes including nutrient source for agriculture. It is also used for purification of contaminated soil, water and air. SMS has also been found to be potential for reclamation and improvement of soil health.

1. Use of SMS for Organic Farming

In present scenario of agriculture, every one interested in availability of non-residual food sources. The organic farming is the only solution for supplying non-residual sources. Spent Mushroom Substrate, used as a nutrient base for development of organic farming in the larger scale. It supplies all the necessary nutrients including NPK for the growth and development plants.

2. SMS as organic fertilizer

Spent mushroom substrate is still nutrient-rich and contains about 80% of the total nitrogen in bound form with high molecular weight fractions of lignin and humic substances. Though, the release of nitrogen from SMS is quite slow, which can be enhanced by the addition some easily available form of nitrogen. Conversion of SMS into an organic mineral fertilizer is an alternative way of using SMS for soil amelioration and plant nutrition.

The organic mineral fertilizers are prepared with three different formulae, having 2, 7 and 10% nitrogen, each supplemented with 2% phosphorus and 2% potash. These formulations have been found to be good for spinach production at large scale.

3. Bioremediation of contaminated soil

The uncontrolled release of industrial waste is the good source for chemical pollution of the environment. When SMS is released in the environment, the chemicals are degraded at the varying rate depending upon the prevailing physical and chemical conditions. The SMS has the ability to chemically absorb the organic and inorganic pollutants. Moreover, it contains a diverse category of microbes having capabilities of biological breakdown of the organic xenobiotic compounds present in the soil and water.

The bioremediation systems are normally of two types. The first type is dependant on the catabolic activities of the indigenous microorganisms present in the contamination site. This activity can be optimized by optimum providing for growth and biodegradation. The second is seeding the contaminated site with pre-tested microorganisms having high catabolizing activity.

The mixing of pentachlorophenol (PCP) contaminated soil with aliquots of spent sawdust cultures of shiitake mushroom, supplemented with nutrient solution of glucose, thiamine and mineral salt result in disappearance of about 44.4-60.5 percent of PCP in 21 days of the incubation. Reduction in PCP does not have any relation with residual polyphenoloxidase activity in the inoculum. SMS has also the decontamination potential of land sites used for disposal of hazardous wastes. SMS is full of microbes especially actinmycetes (*Streptomyces* spp. and *Thermonospora* spp.) having strong PCP catabolising capabilities.

4. Bioremediation of contaminated water

The sulphur deposits from nearby coal mines after mixing with rain water, forms sulphuric acid, leading to acid mine drainage. This acid mine drainage creates unfavourable conditions for aquatic flora and fauna. Such coal mine damage can be well treated with SMS. The water coming from the wetland has high pH. Use of SMS cleans the water to be used for various purposes. SMS is found to stabilize the abandoned coal mine sites, pipeline construction site, etc.

5. Reclamation of Soil

Addition of spent mushroom substrate in the nutrient deficient soil leads to improvement of soil health by improving the texture, water holding capacity and nutrient status. Simultaneously it reduces thermal conductance, bulk density and water stable aggregates of soil. Its amendments in soil increases both pH as well as organic carbon contents. Addition of only 5% SMS can provide the phosphorus and potassium requirement of soil for plants, whereas to fulfill nitrogen requirement, there is a need of 25% SMS for amendment. Reclaimed soil is suitable for raising vegetables.

6. Nursery and horticultural crop raising

Spent mushroom substrate either partial or complete may be used as substitute of growing media for nurseries of flowers, vegetables, fruits, saplings, ornamental shrubs and other horticultural plants of economic importance. Being rich in NPK, it acts as a rich source for growing vegetables like cucumber, tomato, broccoli, tulip, cauliflower, pepper, spinach, etc. tomato grown in SMS based nutrients have more ascorbic acid.

7. Greenhouse crop raising

The application of spent mushroom substrate is not limited to the growing vegetables in open field, but also in the greenhouse grown crops. The SMS and speedel medium are more nutritious than the normal greenhouse soil.

8. SMS for disease management

As we know that SMS has a unique chemical composition and it is full of microflora, including actinomycetes, bacteria and fungi. These microflora exert some antagonistic activities against pathogen surviving and multiplying in the soil ecosystem. The organic amendment of soil with SMS has been found to be helpful in restricting the root-knot nematode (*Meloidogyne incognita*) infestation in tomato crop. Application of SMS has shown the higher efficacy than the carbofuran @ 92 kg a.i./ha. In *in vitro* conditions, an aqueous extract of SMS found inhibiting the conidial germination of *Venturia enaequalis*, the cause of apple scab. These inhibitory effects of SMS remain protected even after the autoclaving of the SMS.

9. SMS for re-growing mushrooms

Production for second crop of mushroom has been advocated out of SMS. Second crop mushrooms can be taken with the addition of nutrients in the SMS. The shiitake SMS supplemented with 10% wheat bran and 10% millet could be utilized for re-cropping of *Pleurotus sajor-caju*. Before cropping, grinding, air drying, supplementation, pasteurization is necessary. Even higher yield up to 79% conversion is achieved if it amended with 12% soybean and 17% calcium carbonate.

10. SMS as Animal feed

Normally the domestic animals are fed with wheat straw, paddy straw as staples with concentrates including mustard seed and cake. The lignin which is generally inaccessible to the animals from the fresh straw is broken down into simple lignin compound after action of mycelium while cultivation of mushrooms. The protein quality of SMS is also improved. SMS from oyster mushroom cultivation has more digestibility than other SMS.

11. SMS for biogas production

The anaerobic fermentation of animal manures, domestic wastes and municipal garbage is carried out for production of methane. SMS recycled as animal feed comes again for biogas production in the form of dung. Dung is generally used for the production of biogas. Generally SMS recycled from oyster mushroom is used as animal feed. Since, the SMS has high nitrogen content which circulate to the dung, making high yield of gas in the biogas plant.

Chapter 14

ECONOMICS OF MUSHROOM CULTIVATION

As with any business venture, properly accessing the feasibility of a mushroom based enterprise requires analyzing the costs as well as potential income. Several of the major expenses include the cost of compost material, the cost of mushroom spawn and other supplies, and the cost of labour. The input 'labour' has been identifies labour as the most significant investment in the mushroom enterprise.

Capital is required for non-recurring expenditures *viz.* building or thatched shed, permanent platforms, equipments, etc. If money has to be borrowed, interest is another cost that would be included in capital requirement.

Buying ready-spawned compost bypasses expensive capital investment in machinery and buildings necessary for the composting. Readymade compost can then be used for growing mushroom. Some growers buy, fully prepared compost, which is ready for casing. Commercially available compost is more expensive but can yield particularly well, increasing cash flow and retained income. However, it will also require a higher level of working capital. The choice of system will ultimately reflect availability of capital, facilities and site restrictions and can be decided only after the course of a detailed financial analysis.

In all cases some working capital is needed to grow the mushrooms. It is often assumed that 'family' (unpaid) labour can carry out some of the production and harvesting operations. While this may be true on a small scale, the crop is relatively labour intensive, therefore, additional labour is almost certainly required for cultivation.

It has been observed that the employment opportunity has been squeezed in the Government sector due to scarcity of funds as well as imposition of restriction by different financial agencies. It is indicated that the financial reforms reduced the average financial growth rate as well as decreased the employment opportunities. Mushroom enterprise for present scenario could be the best deal for self-employed group of people, which can easily be established in the rural areas.

Presently most of the nationalized as well as private banks, NABARD (National Bank for Agriculture and Rural Development), Co-operative banks and Cooperative societies are providing loan for establishment of different projects including mushroom cultivation.

As the enterprise of mushroom cultivation is dependant on agriculture and allied sector, we have to depend upon the agricultural produce of the local area and one has to create awareness among the people of that locality for a particular produce to make available for easy cultivation of mushroom.

Economics for production of one kg grain spawn

a) Cost of cultivation of 1 kg spawn production in polypropylene bag.

	Particulars	Cost (Rs.)
1.	Cost of 700 g wheat grain	7.0
2.	Polypropylene bag, 200 gauze, 1 kg capacity @ Rs. 0.8/bag	0.8
3.	Polypropylene neck	0.5
4.	Non-absorbent cotton for plug	1.00
5.	Butter paper and rubber band	0.50
6.	Inoculum from mother spawn	1.00
7.	Energy cost for boiling, sterilization, incubation, etc.	1.00

8.	Labour cost	1.50
9.	Depreciation on machinery/building	1.00
10.	Interest on capital	1.00
11.	Literature/consultancy	0.50
12.	Chemicals	1.00
	Total	**16.30**
b)	selling price of 1 kg spawn bag	**40.00**
c)	Profit per kg spawn	**23.70**

Economics for commercial cultivation of button mushrooms

Land is required for composting yard, pasteurization chamber, pasteurization room for casing material, spawning room, insulated cropping rooms, broiler room; air conditioned unit, store room, packing room, staff room, etc. The water storage facilities are however, constructed on roof. For these works approximately 16, 000 sq. feet (or 1487 sqm) of land is required.

FIXED COST

	Particulars	Amount (Rs.)
A.	Land procurement and development @ Rs. 750/ sqm.	**11,15,242**
B.	Infrastructure facilities	
i)	Construction of one composting yard (70′ x 40′ x 20′) @ Rs. 150/sq.ft	4,20,000
ii)	Construction of 2 units insulated pasteurization chamber (36′ x 9′ x 12′) @ Rs. 300/sq.ft	1,94,400
iii)	Construction of pasteurization room for casing material ((15′ x 12′ x 12′) @ Rs. 300/sq.ft	54,000

iv)	Construction of underground service room (12′ x 12′ x 12′) @ Rs. 225/sq.ft	32,400
v)	Construction of spawning room (20′ x 12′ x 12′) @ Rs. 225/sq.ft	54,000
vi)	Construction of boiler room (20′ x 12′ x 12′) @ Rs. 225/sq.ft	54,000
vii)	Construction of overhead water tank (25000 ltr capacity) @ Rs. 3/ltr	75,000
viii)	Construction of 12 insulated cropping rooms (35′ x 25′ x 12′) @ Rs. 300/sq.ft	31,50,000
ix)	Construction of A/C unit (15′ x 15′ x 12′) @ Rs. 225/sq.ft	50,625
x)	Construction of Staff room (30′ x 12′ x 12′) @ Rs. 225/sq.ft	81,000
xi)	Mushroom packing room (15′ x 12′ x 12′) @ Rs. 225/sq.ft	40,500
xii)	Store room (15′ x 12′ x 12′) @ Rs. 225/sq.ft	40,500
	TOTAL FIXED COST 'B'	**42,46,425**

C.	**Machinery**	
i)	Central A/C facility consisting of one 100 tonne capacity	25,00,000
ii)	Boiler (300 kg evaporation/hr capacity)	2,50,000
iii)	Generator 50 KVA capacity	4,00,000
iv)	Under stack blowers	30,000
v)	Iron racks in the growing room and iron grating in the chamber	4,50,000
vi)	Compost tuner, mixer and filling line	1,75,000
vii)	Electricity and other installations	1,50,000

viii)	Sprayers, buckets, forks, harvesting knives, rubber pipes, weighing machines, etc.	1,00,000
	Total Fixed Cost C	40,55,000
	Total Fixed (A + B + C)	**94,16,667**

VARIABLE COSTS

A.	Raw Materials and operational expenses	
i)	Wheat straw 400 tonnes @ Rs. 3000/ tonne	12,00,000
ii)	Chicken manure 225 tonnes @ Rs. 350/ tonne	78,750
iii)	Casing soil procurement and its treatment for 450 m^3 @ Rs. 100/ m^3	45,000
iv)	Spawn 18,000 bags @ Rs. 12/bag	2,16,000
v)	Gypsum, Urea, Pesticides	50,000
vi)	Electricity, fuel, and water charges	3,00,000
vii)	Miscellaneous expenses	15,000
viii)	Risk factor @ 4% for total raw material	76,190
	Total variable cost A	**19,80,940**
B.	Manpower requirement and wages	
i)	One manager @ Rs. 6000/PM for one year	72,000
ii)	One typist cum secretarial assistant @ Rs. 3500/ PM for one year	42,000
iii)	One Jr. engineer to look after the machinery work @ Rs. 4500/PM for one year	52,000
iv)	Total man-days involved in compost preparation and spawning. Total 50 (40 man days/outing) outings in a year and wages involved at Rs. 80/man-day.	1,60,000

v)	Labour requirement in the production unit at 70 man days/crop/room, i.e. 50 crops (70 x 50 man-days) at 100% capacity utilization, @ Rs. 80/ man-day.	2,80,000
	Total variable cost B	**6,06,000**
C.	**Interest and depreciation**	
i)	On building, for the cost of Rs. 11,15,242/= @ 2.5% depreciation plus 12.5% interest	1,67,286
ii)	On land (opportunity cost) for the cost of Rs.11,15,242/=(@ 10% interest). Since land receives an annual premium (appreciation) @ not less than 10%, hence the interest will be waved off	Nil
iii)	On machinery cost of Rs. 40,55,000/= @ 10% depreciation and 12.5% interest	9,12,375
	Total variable cost C	**10,79,661**
D.	**Cost of production and return per annum**	
i)	Raw materials (**Variable cost A)**	19,80,940
ii)	Wages and Salary (**Variable cost B)**	6,06,000
iii)	Interest and depreciation **(Variable cost C)**	10,79,661
	Total variable Cost (A + B+ C)	**36,66,601**
iv)	Total production of 144 tonnes of fresh mushroom at 100% capacity utilization and 18% conversion of compost from 4 crops of 90 days (15 for spawning, 15 days for casing, 4 days for primordial initiation and 21 days for first crop, 32 days for consecutive 8 flushes of 7 days each) each	
	Cost of production per kg of fresh mushroom (Rs.)	25.46
	Income from sale of mushrooms at Rs. 60/kg	86,40,000
	Net profit/year	**49,73,399**

Economics for cultivation of white button mushrooms as a small-scale industry (for an area of 100 sq. m).

A.	**Fixed cost**	**Amount (Rs.)**
i)	Land is already with cultivator	
ii)	Cost of construction of mud house of 20 x 5 m @ Rs. 300/sq.m	30,000
iii)	Cost of racks made from bamboo or sticks	5,000
iv)	Cost of equipments including sprayers, buckets, thermometers, etc.	5,000
		40,000
B.	**Variable costs**	
i)	Wheat straw 40 quintal @ Rs. 150/- q	6,000
ii)	Poultry manure 1000 kg @ Rs. 40/-q	400
iii)	Fertilizers	
a)	Calcium ammonium nitrate 50 kg @ Rs. 3/- kg	150
b)	Urea 30 kg @ Rs. 4/kg	120
c)	Superphosphate 20 kg @ Rs. 2/ kg	40
d)	Potash 20 kg @ Rs. 3/kg	60
iv)	Molasses or Wheat bran	300
v)	Pesticides	200
vi)	Spawn 200 bottles (250 g/bottle) @ Rs. 20/ bottle	4,000
vii)	Casing soil preparation charges	500
viii)	Two labourers for 4 months @ Rs. 2100/pm	16,800
ix)	28 man days for compost preparation and filling for trays, etc. @ Rs. 70/man day	784
x)	Miscellaneous expenses	1,500
	Total	**24,454**

C.	**Depreciation and Interest**	
	Depreciation on fixed cost (A) @ 5% for mud house and interest @ 12% for minimum of 4 months	3,200
	Total expenditure (B+C)	**27,654**
	Anticipated yield at 18% of conversion of composting material	720 Kg
	Cost of production per kg (Rs.)	38.41
	Income from sale at Rs. 60/ kg	43,200
	Net income in 4 months	**15,546**

ECONOMICS FOR CULTIVATION OF OYSTER MUSHROOM

A.	**Fixed cost**	**Amount (Rs.)**
i)	Land is already with cultivator	
ii)	Cost of construction of thatched house of 60′ x 20′ x 12′ @ Rs. 10/sq.feet	12,000
iii)	Cost of racks made from bamboo or sticks	5,000
iv)	Cost of equipments including sprayers, buckets, thermometers, etc.	5,000
v)	Cost of two drums/tubs	1,000
	Total	**23,000**
B.	**Variable cost**	
i)	Cost of 5 quintals of wheat straw @ Rs. 300/q	1,500
ii)	Cost of polythene bags	500
iii)	Cost of 100 spawn bags (400 g/bag) @ Rs. 24/bag	2,400
iv)	Cost of pesticides	300
v)	Miscellaneous charges	1,000
	Total	**5,700**

C. Depreciation and interest

Depreciation @ 5% and Interest (at least for 2 months) @ 12% on fixed cost	1,610
Total expenditure (B+C)	**7,310**
Anticipated yield @ 63 conversion efficiency of substrate	315 kg
Cost of production per kg (Rs.)	23.2
Income from sale @ Rs. 60/ kg	18,900
Net profit	**11,590**

Economics for cultivation of paddy straw mushroom

As we have seen, for the cultivation paddy straw mushroom, there is no requirement of any permanent structure, etc. therefore, the **fixed costs** in the paddy straw cultivation is almost **nil**.

VARIABLE COSTS

S.N.	Particulars	Amount (Rs.)
(i)	Cost of 30 quintals of paddy straw @ Rs. 50/ q	1,500
(ii)	Cost of 170 bags (400 g/bag) spawn @ Rs. 24/bag	4,080
(iii)	Cost of pesticides	150
(iv)	Besan 5 kg @ Rs. 40/ kg	200
(v)	Labour 50 man days @ Rs. 70/	3,500
(vi)	Miscellaneous charges	1,000
	Total	**10,430**
	Anticipated yield @ 12% conversion of substrate on dry weight basis	360 kg
	Cost of production Rs. per kg	28.97
	Income from sale @ Rs. 40/ kg	14,400
	Net profit	**3,970**

Economics for shiitake Mushroom Production

The economic considerations are made inoculating 4,000 logs per year for a 15-year planning period.

Attributes for calculating fixed cost or Capital Expenses

(i) Building for indoor works

(ii) Logs

(iii) Soak tank concrete

(iv) Laying yard (Shade cloth, wooden poles, and steel cable (to hold up shade cloth)

(v) Tractor

(vi) Refrigerator

(vii) Digital Scale

(viii) Wax or plastic foam plugs

(ix) Aluminum identification tags and staples

Attributes for calculating Operating Expenses

i) Log Covers made up of plastic and fabric

ii) Tools and Supplies

a) Sawdust spawn inoculation tool

b) Staple gun

c) Log drilling stand (sawbuck)

d) Electric drill (high speed)

e) Drill bits

f) Electric extension cord

g) Wax melting pots

h) Wax baster or brush

i) Water hose/ sprinkler head

j) Scale for weighing logs (to determine moisture content)

k) Mushroom picking/ storage baskets

l) Laying yards maintenance materials

m) Steel racks for carrying/soaking logs

n) Office supplies

o) Tractor operation / maintenance

iii) Utilities

a) Outdoor - water/electricity

b) Indoor (optional indoor fruiting during winter) - water/electricity/heat

iv) Advertising

v) Shipping

a) Packaging and labels

b) Transportation

Interest on borrowed money

Chapter 15

PEST AND DISEASE MANAGEMENT

Mushrooms are attacked by a number of fungi, bacteria, mites, nematodes and viruses. Each mushroom species in a specific environment has a different pest complex. Even a little carelessness in the cultivation of Mushrooms can cause havoc in the Mushroom industry. Improper pasteurization of compost and casing becomes major source of infection of fungi and bacteria. Once the disease has introduced. It attracts mushroom flies, a common pest among many cultivated mushrooms, on the smell of decaying mushroom substrates. A predatory nematode attacks this fly in its larval form. The secondary infection is generally carried out by the different agencies such as air, water and workers working in the mushroom house. Hence a chain of pests develops, which attack mushroom crop causing huge losses.

A number of insects, fungi, bacteria, viruses, nematodes and other parasites affect the mushroom production. A complete hygienic conditions, either by chemicals like 5% centrabride formaline, 0.05% melathion or 0.03% Newran, or some others means suggested in the following section, should be maintained.

DISEASES OF MUSHROOM

There are following important diseases of Mushroom, which occur frequently:

Dry bubble

This disease is caused by *Verticillium fungicola* Gams. The pathogen produces several recognizable symptoms. An often-encountered symptom is dry bubble, the disrupted growth of a pin or buttons into a ball-like mass, generally the size of a grape or a bit longer. Another common symptom is stem (stipe) "blow-out"- the stem shatters and the cap tilts slightly. Many times the fungus pathogen can be seen fruiting on infected tissue of the cap where it appears as a grayish hue in the infected area. A less recognized form of the disease is called flyspeck or spot. Spots develop when the pathogen infects a single site on the cap, which developed into a discrete spot.

The pathogen arrives on the cap in splashed water or on hands or tools of a picker. The fungus usually enters a growing room via casing that is not adequately treated, on infested tools and picking instruments, and dust particles in the air; it can also be carried in by flies and on the clothing of workers. When present, the severity of disease may range from a trace to a yield loss of as much as 70 percent of the crop. Even if yield is not reduced, infected mushrooms are spotted or deformed and thus penalized in terms of quality.

Management

Proper pasteurization of compost as well as casing soil and control of insect vectors like flies, mites, nematodes can escape the disease. Dithane Z-78, Sporgon, and Topsin M may be applied in the stage of infection.

Soft Mildew or Cobweb

A fungus *Dactylium dendroides* causes this disease, where a fluffy, white cobweb like growth occurs over the surface of casing soil. Initially it is white but later changes to pink in colour. The attacked mushroom buttons appear white on the surface but they are rotten from inside. The chief source of infection for this disease is soil humidity, particularly when grown in wet trays.

Management: A good ventilation and prevention of excess humidity in the mushroom house can check the further development of disease. However, individual patches may be treated with Dithane-Z 78 @ 0.2%. Before starting with new crop, mushroom house should be disinfected with farmalin solution.

Brown Plaster Mould

This disease usually occurs in spawning and cropping trays. First it is white with a cloudy appearance, later the colour changes to brown.

The disease is caused by *Populospora byssina.* This fungus is very rapid in growth. Infection occurs through the compost, if it is too wet while preparation. Moreover, high temperature between 28-32 ^{0}C during spawn running stage and above to 18 ^{0}C during cropping time favours the disease.

Management: Compost should be prepared in a proper way, going through schedule provided for composting. It should not be left too wet. Temperature should be kept below the required temperature for fungus. Compost house should be disinfected with 2% formalin spray.

White Plaster Mould

The white plaster mould is caused by *Scopulariopsis fumicola.* The disease closely resembles to brown plaster mould in the initial stages but in the later it changes from white to delicate pink shade.

This disease is also induced from the improper compost. Too wet compost attracts the fungus. The disease can be managed in way suggested for brown plaster mould.

Olive Green Mould

Chaetomium olivaceaum (the causal agent) fungus appears in the in the compost after spawning but before casing. It the compost is prepared in improper way and the pasteurization is inadequate, then the fungus may appear before spawning too. The fungus is characterized by the production of white growth in the initial stage, but shortly changing to olive green colour.

Management: Sufficient air should be introduced without increasing the temperature above 60 ^{0}C during pasteurization. It is also advisable to spray a @ 0.05% solution of benlate to the mushroom trays.

Inky Cap

The disease is caused by *Coprinus lagopus* and *C. comatus* and characterized by the appearance of a long slender stalk with a small thin cap. This slender stalk with cap in due course of time dissolves in to black inky liquid.

This disease is favoured by the ammonia, which might have left in the compost while peak heating. The appearance of the black inky liquid in the mushroom trays, denote this fact and confirm the presence of ammonia in the compost.

Management: Disease can be escaped, if while preparation of compost ammonia has completely being removed. Also if profuse growth occurs, compost from the spawned trays should be re-pasteurized as above. The re-pasteurized compost should be filled in trays, spawned, and cased again.

Green Mould

This disease is caused by *Trichoderma viride*, a fungus that is mostly used as a bio-control for several fungi. The fungus appears in the form of green patches on the trays generally after spawning and casing. Since, the disease occurs in the spawned trays, the spawn run is affected. It also appears after casing hence it also affects pinhead formation as well as the development of emerged pinheads.

Management: Since, the fungus is favoured by improperly pasteurized compost and high humidity in mushroom house;

therefore humidity should be lowered in the current cropping stage. Application of benlate @ 0.05% is also recommended. For next crop compost should be properly pasteurized before filling into the trays.

Brown spot disease

The disease is caused by *Verticillium malthousei* or *V. psiallistae* and characterized by the formation of small light brown spots on the cap. These small spots form larger irregular area upon coalescing. In the severe condition mushroom cap become malformed, shrink and leathery.

Management: Since, the disease is correlated with improperly pasteurized casing material, it should be done in next crop. It is also advised to maintain proper temperature and ventilation during growth time. Disease may be managed by the spraying of 0.2% dithane Z-78 at the time of casing, pinhead formation and during the flushes of harvesting.

Bacterial Blotch

Bacterial blotch caused by *Pseudomonas fluorescens*, biotype Va (*P. tolaasii*). Blotch often appears on the sides and lower edges of the mushrooms. Typically, blotch has a bronze color, but cap splitting or a yellowish cast on the cap may be signs of the disease. On very rare occasions the disease can cause the stipe of the mushroom to turn back and shatter, and the cap may also blacken. The pathogen needs a wet cap for 6 to 12 hours for disease to occur.

Improperly pasteurized casing soil is conducive factor for the development bacterium. The bacterium is transmitted through water, flies, mites and nematodes, causing secondary infection.

Management: Casing soil should be properly pasteurized. Adequate ventilation should be maintained. It is advisable to used chlorinated water, which reduces the disease. Before the pinhead formation, a spray of teramycin @ 9 mg/ft^2 is recommended.

Nematodes Diseases

There are three kinds of nematode found in the mushroom house. The nematodes develop in the improperly prepared compost and casing material. The plant parasitic nematodes bear stylet, which help puncturing the mycelium and sucking its contents. They puncture mycelium at several places and suck contents of mycelium. As the nematodes keep on eating the mycelium, it starts disappearing from the mushrooms trays. Nematodes also feed on the pinhead and mushroom caps. The after attack turns into brown in colour, becomes watery, and stipe becomes stunted. There are several species including *Dytelenchus mecelophagus* and *Aphalenchoides composticola,* which feed on mushrooms.

In addition to the feeding on mycelium, nematodes also act as the carrier for several viral, bacterial (e.g. *Pseudomonsas tolassi*) and fungal diseases. They sometimes excrete toxic materials being harmful for mushroom crop.

Control: Nematodes can be controlled by adding 40 ml of nemagon in 10 litre of water for 300 kg of wheat straw while composting.

Viral Diseases

So far none of the virus has been reported from India causing infection mushroom crop. However, there are reports from other countries for the presence of viral particles in the mushroom fruiting body causing a disease syndrome. This syndrome include, brown fruiting body, watery stipe, and dieback, etc. the diseased mushroom become leathery, shriveled and changes colour to brown in the normal dry conditions. Stipe becomes watery and turns to grey in colour in the humid conditions. Sometimes an etiolated growth of stipe with tilted fruit is also observed, which looks alike drumstick. However, the infection of virus may remain latent. According to reports, off white varieties of mushroom are less prone to disease than the white varieties.

The transmission of viruses occurs through phorid larvae and tarsonemus mites. The mushroom species including *Lacaria laccata* have been reported to be the reservoir of the mushroom viruses.

Management: Mushroom house should be heat sterilized before starting the crop. Disinfection of mushroom with 4% molybdate solution is also recommended. Phorids may be controlled with suitable systemic insecticide, to check further spread of disease.

INSECT PESTS

Sciarids and Phorids

Two insect pests of commercial mushrooms are considered: *Lycoriella mali* (Fitch), a **Sciarid** or darkwinged fungus gnat, and *Megaselia halterata* (Wood), a Phorid or humpbacked fly. *Lycoriella mali* is commonly referred to as the Sciarid or "big fly," passes through four stages in its life cycle *viz.* egg, larva, pupa, and adult. The latter three stages are easily seen without a hand lens, but magnificent is required to see the eggs.

Sciarid eggs are smooth, oval, whitish and translucent; as they develop, the distinctive black head capsule of the larva becomes visible. Phorid eggs are also whitish, but oblong and concave. Eggs of both flies are deposited in compost.

Sciarid larvae are translucent white, segmented, and have a black-headed capsule. Four full-grown larvae laid end-to-end cover a span of 1 inch. Phorid larvae are also whitish and segmented, but do not have the external black head capsule. Phorid larvae resemble the house fly maggot- blunt on the rear end and tapered on the front end, but considerably smaller. Six mature Phorid larvae laid end-to-end measure 1 inch. Sciarid larvae are often found inside the mushroom stem while Phorid larvae remain in the compost.

The pupae of the two flies differ considerably: Sciarid pupae have barely visible legs, which hang loosely. Phorid pupae have a "cocoon-like" appearance. Both species pupate in the compost.

The adult stages of the two species differ in several respects. Female Sciarid flies are larger and more rotund than the humpbacked adult Phorid and have antennae that are easily seen.

Antennae on the Phorids are difficult to be seen. While identification one should cautious because Sciarid males are about the same size as an adult Phorid. Form of adult movement is also a clue to identification; Phorids move with quick jerky movements, while Sciarid movement is less erratic.

Besides knowing what flies are present, growers should know something about the development time needed for an insect to evolve from one stage to another; this varies greatly with temperature for each insect. To review a few general characteristics of common insects in mushroom growing operations:

1) Egg and Larvae development time is longer for the Sciarid;
2) Pupae-development time is longer for the Phorid; and
3) Total generation time is longer for the Phorid.

Both species exhibit behavioral patterns significant to a management programme. Sciarids enter the mushroom house during Phase II cool down as soon as the air temperature drops to below 43 ^{0}C; they lay their eggs on the unspawned compost. Phorids infest the house only after the beds or trays have been spawned, the greater numbers entering 4 or 5 days after spawning and to a lesser extent after casing. Phorids lay eggs only on spawned compost. Lastly, adults of both species are attracted to light, especially fluorescent black light, with the exception of the Sciarid adult male, which is not attracted by light.

Mites

Mites are generally present in the mushroom house. Mites directly do not harm the mushroom crop; rather they are the carrier disease propagules. They generally feed on mycelium of the fungus, keep it within them, after sometimes make hole into mushroom cap and vomit the propagule out into the cap. Propagule after entering into the cap becomes active and cause infection.

Management: Mites of *Phylloxera* spp. can be managed by spraying of spawn trays with 7 ml of 50% E.C. malathion in 10 litres of water few days after spawning and 2 days before casing.

ABNORMALITIES IN MUSHROOMS

Rose comb

Rose comb in mushrooms is characterized by the malformation of pileus. The gill which form at the under side of the mushroom cap, in malformation they form on the top of the pileus. Since, the gills are formed on the top, mushrooms appear as the colour of spores through gills. Gills are also malformed appearing as irregular in shape.

Management: Rose comb is correlated with the traces of mineral oil, which generally come through the sprayers used for watering the beds. Therefore, it is advised to clean the spraying machines before use. Also avoid mineral oils in the mushroom house, as its fumes cause such abnormalities in the mushrooms.

Elongated mushroom stalks

In this abnormality, mushroom stipes becomes unnatural long. Conversely, the mushroom pileus becomes small. The cap may take irregular shape also. This abnormality has been correlated with the improper ventilation in the mushroom house.

Mushroom cracking

Mushroom cracking is also a kind of deformity, which occurs due to the fluctuation in the temperature in the mushroom house. The temperature fluctuations cause, sometimes, dryness and excess humid environment. Maintaining a uniform temperature may manage the abnormality.

Premature opening of caps

High temperature in the mushroom house induces the early opening of pileus. This unnatural opening of pileus may be attributed to high concentration of carbon dioxide developed due to high temperature gradient.

INTEGRATED PEST MANAGEMENT

IPM is a least-toxic approach for managing any pest in question. The integrated management of a pest is accomplished by altering the environment to be disadvantageous for the pest. In order to achieve full control over the pest population, the pest has to be identified and some of the questions including "what pests are active", "how many pests are involved" to lessen the profits. If the proper understanding of the life cycle of each problematic organism in known, it can help in measuring the pest population at right of the developmental phase, bringing the population below the economically damaging threshold level. Insect pests are subjected to several enemies and can be controlled in easy way.

Here are some examples of non-chemical methods used to control typical pests in the production of white button mushrooms. Mushroom flies, a common pest among many cultivated mushrooms, are attracted to the smell of decaying vegetation such as mushroom substrates. Screening the mushroom house ventilation system will keep adult flies out. Double doors and positive atmospheric pressure within the structure also prevent flies from entering the house.

Biocontrol is another option for several mushroom pests including the sciarid flies. A predatory nematode attacks this fly in its larval form. Therefore, this nematode can be added to the composting substrate to prevent infestation.

PREVENTIVE MEASURES

A range of diseases and insect pests can attack mushrooms. Prevention is the best way of controlling pests and diseases in mushroom crop. It is absolutely essential to keep the mushroom house clean. Use clean tools and equipment, clean compost, and pure, uncontaminated spawn from a reliable source. If any sign of a pest or disease appears, it should be managed immediately. There are some of the prophylactic measures, which may serve to keep crop, devoid of pests and disease.

Safe Sprays

It is difficult to find any safe pest or disease sprays to use on mushrooms when they are close to harvest. One possibility might be herbal sprays, as they are relatively safe and even after spray mushrooms can be eaten immediately. Though many herbal sprays are mild in their effect on pests or diseases, they may be good enough in the indoor situation. It is advised to growers, that before using these sprays, one must test the volume and strength of spray with the suitability of their own crop in special niche, as there might be some possible side affects.

Garlic Spray

Garlic has the repellent properties for insects; it also bears certain fungicidal properties. One spray has almost 10 days residual affect on the crop.

Preparation: Crush 90g of garlic cloves; pour 10 ml of non-aromatic oil over them; soak for 48 hrs; add to this, a soap solution (about 10 gm of soap in 500ml water); leave for one day and then strain; store in sealed containers; use 1 part mixture to 100 parts water or less depending on situation and severity of insect population.

In addition to the garlic spray, Lavender spray can also be used against flies, which are major problem on mushroom crop.

Chapter 16
MARKETING MUSHROOMS

Marketing of fresh mushrooms is always done in the nearby city, especially Delhi, Bombay, Madras, Chandigarh and others. Most of the produce from big commercial farms is canned in brine and exported to destinations outside India, especially USA. The quality of the mushrooms exported is excellent as most of the big commercial farms are growing hybrid strains of *A. bisporus* made available to them by multinational spawn companies like Sylvan, Amycel and others. The mushrooms are blanched and preserved in brine in large containers for shipment to distant destinations in containers, and are repacked at the final destination to suit local markets. A quantity is freeze-dried by the producer and exported at a good price. The mushrooms for freeze-drying require to be picked as smaller buttons, hence reduced yields.

There are two main types of mushroom growers in India, seasonal growers and round the year growers. Both grow white button mushroom for the domestic market and export. The seasonal button mushroom growers are confined to temperate areas such as Himachal Pradesh, Jammu and Kashmir, hilly regions of Uttar Pradesh, hilly regions of Tamil Nadu and North Eastern hilly regions where growers take 2-3 crops of button mushrooms in a year. Also included in the seasonal growers are the growers from North Western plains of India who grow one winter crop of button mushrooms and sell it fresh.

Consumption and Price

The per capita consumption is 15-20g only and by increasing the per capita consumption to 100g, we should be able to market 100000 tonnes of mushrooms within the country. There are positive signs emanating from the consumer in India, and today cultivated mushrooms are available in all common vegetable shops, grocery stores and departmental stores in small and big towns in India. Mushrooms are more relevant to predominantly vegetarian India.

According to a survey report of Mac-Plan India, about 60% of the total consumption, is consumed in Northern Region, and 20% in the Eastern Region. They have also observed that consumers with income ranging between Rs.10, 000 and Rs. 35,000/- were the major consumers. However, the supply and price trend of mushrooms are inconsistent affecting consumer demand. It is estimated that demand-supply gap by 2006 will be around 1,26,000 MT/year which clearly denotes a vast scope to expand mushroom enterprises.

The retail price of fresh mushrooms in Indian ranges from Rs. 50/- to Rs.100/- per kg, depending upon the season. In summer months the prices are higher than in winter months due to the high cost of electricity for cooling. Prices are lower in winter due to arrival of mushrooms into the market from seasonal growers. Prices of mushrooms in the marriage season go as high as Rs.150.00 per kg for a short period due to greater demand.

India being a tropical country, fresh marketing is at a premium, except for a brief winter period. The commonly used packaging is the polythene bag. Most of the mushrooms sold in fresh markets are treated with potassium metabisulphite due to market demand as mushrooms become extra white after the treatment and the casing adhering is also removed.

Export Oriented Units (EOU)

The all-season growers are scattered all over the country. The large/export oriented units (EOU) are located in Punjab (near Chandigarh), Dehradun (Near Massoorie hills), Gurgaon (near Delhi), Hyderabad (South India), Madras (South India), Ooty hills (South India), Pune (near Bombay), Paonta Sahib/Nalagarh (HP) and at Goa (Western India). These large units have the growing capacity in the range of 200 to 5000 tonnes per annum. **M/s Agro Dutch located in Punjab (near Chandigarh) is the largest EOU with installed capacity of about 5000 tonnes.** Some of these units were built with the assistance from foreign machinery sellers/ consultants; but entire air handling/cooling machines are now manufactured and commissioned in India.

Major export/Import Markets

The export market for India is chiefly the USA, with some quantities going to UAE, Russia, The Netherlands, Germany, UK, Switzerland, Denmark, Israel, Sweden and other countries. There is no quota available from EU for India, and Indian exporters have to sell processed mushrooms in the EU with additional taxes levied as per the laws of the EU, which makes it difficult for the Indian exporters to compete in the EU market.

The major importing countries are U.S.A., Canada, Germany, Netherland, France, Switzerland, Japan and Malaysia. Fresh mushrooms are preferred over preserved mushrooms in European Union and American countries. Earlier, major quantity of mushroom was exported in preserved/processed form but of late the scenario is changing.

The entry of India in International market coincides with the crash of prices for mushroom in international markets because of competitors like China. China was the leading exporter. However, U.S.A reduced the strong presence of China in market of U.S.A. due to imposition of anti-dumping duties.

M/s. Ponds established an unit at Ooty for export of packed mushrooms to USA. However, some firms have also come in this business, throwing their export potential towards European union as well as American Countries. Now high tech machinery with environmental control equipment is for processing of mushrooms, which are used to be sent in said countries.

Domestic Market

Mushrooms are marketed as fresh, dried and preserved in all the forms. Majority fresh mushrooms from hilly tract are consumed in plains of Punjab, Haryana and Delhi during October-November and February-May. During Dec-January, the produce of plain areas also arrives in markets. The middlemen and commission agents collect produce from growers and dispatch to the regulated markets/satellite markets or markets outside the state. From consumption trend, it is seen that people of high-income group and restaurants catering them are main buyers of mushrooms.

The Mushroom Business Perspective

Market research and evaluation is perhaps the most challenging part of developing a new enterprise. Before entering into this business challenge, the following must be kept in mind:

- Production should be market driven, which will be decided by potential buyers of volume and prices. Talk to the potential buyers.
- Explore various marketing options: brokers, distributors, farmers' markets, restaurants, grocery stores, food service operations, and co-operatives.
- Consider reselling other growers' mushrooms to offer more variety and larger volume.
- Talk to other producers and perhaps a consultant about production systems.
- Consider buying used equipment to reduce initial capital investment.
- Strike a balance between undercapitalizing and a heavy debt load.

Market Demand

As mentioned earlier, *Agaricus* mushrooms are the most loved mushrooms from the mushroom industry. Large, well-established companies produce virtually all *Agaricus* mushrooms. Their production houses are full of mushrooms in every stage of development. Mushrooms raised in these systems can be sold profitably on the wholesale market. It is very difficult for a beginning grower to compete with these companies at wholesale prices.

Market Research

The goals of market assessment are depend upon the sales volume and gross income of the enterprise. Market research is necessary to analyze the potential profitability and cash flow, and to gather information about potential buyers and competitors for setting a new business.

If one is thinking about starting a commercial mushroom enterprise, begin at the end: where to sell them? One cannot make money out of any business if don't have buyers for your product. These points must be learnt that who buys mushrooms, what kinds they want, and where do they shop. Another function of market research may be to evaluate the competition, which will help determine what market already exists. While study identify any lacunae one could fill.

Market strategies

Market strategies should be explored as per the demand. There could following strategies one can follow them:

- Market the fresh or dried product directly to customers through farmers' markets, over the Internet, through mail-order offerings.

- Add value to the mushroom by creating processed products.
- Wholesale as fresh produce through contract.

Direct marketing

Direct marketing is always profitable for growers. If we sell mushrooms or mushroom products directly to an end user, we will naturally receive a better price than selling to a wholesaler. Direct marketing of mushrooms at local farmers' markets, to restaurants, or in supermarkets is possible in many locations. When competing in local markets, excellent service, top quality, and consistent supply, rather than the lowest price, might win the sale. Sometimes, local chefs are specialized in locally grown mushrooms; hence they buy the known product. In any case, establishing a relationship with the buyer and reliably delivering a quality product are essential for this type of marketing.

Local grocery stores are another potential buyer of fresh mushrooms. However, grocery buyers are always interested in buying dried mushrooms, if supplied.

The wholesaler with an established account creates stiff competition; the small, efficient producer might still have an advantage in some niche markets. For instance, shiitakes grown on logs are generally of higher quality and have a longer shelf life than shiitakes grown on sawdust substrates (the most common mass-production method). Log-grown shiitakes earn prices from three to eight times higher than those grown on sawdust substrates.

Locally-grown oyster mushrooms have an advantage because oysters have a very limited shelf life and are too fragile for transportation.

Wholesale markets

Selling fresh mushrooms to a wholesaler will mean a lower price than if you market directly. However, for growers who choose not to involve themselves in direct sales, there are established wholesale markets for mushrooms. Wholesalers advertise for produce industrial periodicals, televisions, Internet ads etc.

Small-scale commercial production of white button mushrooms and other *Agaricus* varieties is not recommended for the beginners, except on a small scale for direct marketing. A significant capital outlay and a high level of management skills are required to begin production, and at current prices, recovery of the initial investment might not be possible. The market is extremely competitive.

Adding Value to Fresh Mushrooms

Adding value to fresh mushrooms usually means either developing a processed product, such as a sauce, or drying surplus mushrooms for sale in the off-season, when prices are higher. A value-added product can be sold either directly to the consumer or to wholesalers.

Chapter 17

MEDICINAL VALUE OF MUSHROOMS

Medicinal properties have been attributed to mushrooms for thousands of years. Mushroom extracts are widely sold as nutritional supplements and touted as beneficial for health. The antitumour mechanisms of several species of whole mushrooms as well as of polysaccharides isolated from *Lentinus edodes, Schizophyllum commune, Grifola frondosa,* and *Sclerotinia sclerotiorum* are mediated largely by T cells and macrophages. Despite the structural and functional similarities of these glucans, they differ in their effectiveness against specific tumours and in their ability to elicit various cellular responses, particularly cytokine expression and production.

According to some estimates, there are over 38,000 species of mushrooms, out of which about 2,000 have nutritious and edible importance. Of these edible and nutritious ones, 500 are known to possess medicinal properties. However, of these 500 medicinal mushrooms, 300 species are most medically actives ones. To date, about 35 mushroom species have been explored commercially and within these, 20 are cultivated on an industrial scale.

The majority of these cultivated species are both edible and possess medicinal properties; but *Ganoderma lucidum* and *Trametes (Coriolus)* spp. are distinctly inedible and medically active. Researchers have discovered that mushrooms contain antibacterial and other medicinal substances, including anti-tumour compounds called triterpenoids.

Several substances of mushroom origin that express promising antitumour, immune modulating, cardiovascular and hypercholesterolemia, antiviral, antibacterial, and antiparasitic effects have been identified. There are about 200 species of mushrooms that have been found to markedly inhibit the growth of different kinds of tumours. Now, the search for new antitumour substances in mushrooms has become a matter of great significance.

Several antitumour polysaccharides such as hetero-beta-glucans and their protein complexes (e.g., xyloglucans and acidic beta-glucan-containing uronic acid), as well as dietary fibers, lectins, and terpenoids have been isolated from medicinal mushrooms. These polysaccharide antitumour agents have been developed from the fruiting body, mycelia, and culture medium of various medicinal mushrooms including *Lentinus edodes, Ganoderma lucidum, Schizophyllum commune, Trametes versicolor, Inonotus obliquus,* and *Flammulina velutipes*. Both cellular components and secondary metabolites of a large number of mushrooms have been shown to affect the immune system of the host and therefore could be used to treat a variety of disease states.

Background information

Mushrooms have been used medicinally by many cultures. Even Hippocrates prescribed them for healing. Asians have known for many years that shiitake mushrooms have medicinal powers with the ability to lower cholesterol and blood pressure, boost the immune system and inhibit tumour growth. In 1960, a scientist at the University of Michigan discovered that shiitake mushrooms contained an antiviral substance that could stimulate the immune system. Most studies on shiitakes have taken place in Japan where their health benefits have been noted for boosting the immune system to produce more interferon, the body's defense against viral and bacterial infections.

During the past 50 years, several major advancements in medicine came from lower organisms such as moulds, yeast, and mushrooms. The first, **antibiotics** were extracted from fungi. Penicillin, tetracyclene and aureomycin, derived from moulds, were hailed as a wonder drug for infections from communicable diseases. We have also seen a rapid pace of advancement in organ transplant due to Cyclosporin, a drug derived from a fungus that uses insects as its host. Cylcosporin suppresses the immune system of transplant patients hence lowering tissue rejection rates. These lower organisms are used to commercially produce bread, beer, wine, cheese, organic acids, and vitamins - including Vitamin C. The Vitamin C tablet you take may be a by-product of fungal growth.

Mushrooms are valuable health food, low in calories, high in vegetable proteins, chitin, iron, zinc, fiber, essential amino acids, vitamins and minerals. Their legendary effects on promoting **good health and vitality** and increasing adaptive abilities of human bodies, have been supported by recent studies. Studies have also suggested that mushrooms are probiotic in nature, as they help our body strengthen itself, and fight off illness by maintaining physiological homeostasis, thereby restoring our bodies balance and natural resistance to disease.

The compounds those mushroom contain, as described, have been classified as **Host Defense Potentiators** (HDP), which can have **immune system enhancement properties**. That is one of the reasons they are currently used as adjuncts to cancer treatments in Japan and China. "In Japan, Russia, China, and the U.S.A. several different polysaccharide antitumour agents have been developed from the fruiting body, mycelia, and culture medium of various medicinal mushrooms (*Lentinus edodes, Ganoderma lucidum, Schizophyllum commune, Trametes versicolor, Inonotus obliquus, and Flammulina velutipes*).

Both cellular components and secondary metabolites of a large number of mushrooms have been shown to effect the immune system of the host and therefore could be used to treat a variety of disease states."

Mushroom species with medicinal properties

Reishi mushroom (*Ganoderma lucidum*) has been called an "**immune potentiator**" and can increase the production of Interleukin-1 and 2 being inhibitory to the tumour growth. Recent studies have also indicated that Reishi can have a number of other effects including Analgesic, Anti-inflammatory, Antioxidant, Antiviral through inducing interferon production, Lowers blood pressure, Cardiotonic action through lowering serum cholesterol, Expectorant and Antitussive, Liver (Hepatitis)-protecting and detoxifying, Protection against ionizing radiation, Antibacterial, and Anti-HIV activity. Reishi contains calcium, iron and phosphorus as well as vitamins C, D, and B including pantothenic acid, which are essential to nerve function and the adrenal glands.

Ganoderma mushroom works in the treatment of cancer because it helps cleanse the body from toxins and it helps strengthen the immune system. It enhances liver detoxification, thus improving liver function and stimulating the regeneration of liver cells, making it a very important supplement for those who have liver cancer. Ganoderma can be used as a supplement during chemotherapy or radiotherapy to reduce side effects such as fatigue, loss of appetite, hair loss, bone marrow suppression and risk of infection. It can also reduce the toxic and side effects and mitigate the pains during chemotherapy and radiotherapy, in particular to cancer patients at terminal stages for prolonging their lives and improving their living quality.

Shiitake (*Lentinus edodes*) mushroom is called "**Elixir of Life**" has been licensed as an anti-cancer drug by the Japanese FDA. Its extract called Lentinan, has shown some effect on bowel cancer, liver cancer, stomach cancer, ovarian cancer and lung cancer. Lentinan stimulates the production of T lymphocytes and natural killer cells and can potentiate the effect of AZT in the anti-viral treatment of A.I.D.S.

Shiitake is also rich in several anti-oxidants (Selenium, Uric acid, Vitamin A, E, C) as well as Vitamin D. Shiitake mushrooms may also lower blood pressure in those with hypertension, lower serum cholesterol levels, increase libido, stimulate the production of Interferon which has anti-viral effects, and has proven effective against Hepatitis in some cases.

The right species of ***Cordyceps*** can be a powerful stimulant for macrophage activity, strengthening your immune system's ability to fight against bacterial and viral infection. Moreover, it is effective for treatment of high cholesterol, poor libido/impotence, arrhythmia, lung cancer, and chronic kidney failure. It is also reported that Cordyceps causes smooth muscle relaxation. This can make it especially helpful for treating chronic coughs, asthma, and other bronchial conditions.

Tremella fuciformis stimulates the immune system, demonstrates anti-tumour activity, lowers LDL cholesterol, protects the liver, is an anti-inflammatory and slow down the aging process.

Maitake (*Grifola frondosa*) is the latest mycologic superstar, shown to stimulate the immune system and is effective against some tumours.

These species of mushrooms (Shiitake, Reishi, Cordyceps) possess some of the compounds (polysaccharides, nucleosides, triterpeniods, complex starches, etc.), which either alone or in combination improve human immune system as well as aid in neuron transmission, metabolism, hormonal balance and the transport of nutrients and oxygen. They also regulate the development of lymphoid stem cells and other important defense responses, through host-mediated (T cell) immune mechanism.

Most of the immune activation provided by medicinal mushrooms is attributed to a complex carbohydrate that enhances white blood cells that destroy harmful microorganisms and tumour cells.

Complex sugars and their derivatives found in mushrooms are able to stimulate a higher level of cytokine production in human beings. Cytokines are proteins produced by the immune system to facilitate communication between cells. Some common cytokines include interleukins, interferon, natural killer cells (NK cells) activating factors and tumour necrosis factors.

Other mushroom species and their medicinal values

Mushroom	Medicinal property
Agaricus bisporus	Stimulating digestion, curing hypertension
Auricularia polytricha	Strengthening health, helping blood circulation
Boletus edulis	Causing muscles and joints to relax
Coriolus versicolor	Curing chronic diseases
Ganoderma lucidum	Rejuvenating effect, neurasthenia
Griflola forndosa	Strengthening health and resisting diseases
Hericium erinaceus	Beneficial to heart, kidneys, lungs, liver, spleen
Lentinu edodes	Strengthening health and resisting diseases
Pleurotus ostreatus	Causing muscles and joints to relax
Tremella fuciformis	Strengthening health and resisting diseases
Tricholoma mongolicum	Beneficial to stomach and intestines

Growth Hormone Enhancers

Gradual decline in the secretion growth hormone, results in ageing of human beings. Human growth hormone (hGH) is responsible for growth spurts in teenagers; it also restores lost muscle mass and redistributes fat cells in the elderly. After the age of 30 years, circulating growth hormone levels decline. Consequently, body composition shifts from muscle to fat. The human growth hormone has gained popularity as a performance-enhancing drug in professional sports.

The first explanation on Arginine and ornithine (amino acids) as precursors for hGH stimulation, came from book, *Life Extension-A Practical, Scientific Approach,* authored by Durk Pearson and Sandy Shaw in 1987. At least 6 g arginine or 3 g ornithine-are sometimes required for a detectable increase of hGH in the bloodstream, according to the authors.

Glutamine is the most abundant amino acid in the human body, and invaluable in influencing hGH. It affects the muscular protein synthesis and circulating growth hormone levels. A little of only 2 g/day oral glutamine inhaling causes a significant release (more than 400 percent) of circulating growth hormone in the body, thereby enhancing muscle growth and improved athletic performance.

Creatine

Creatine is a naturally occurring amino acid (protein building block) found in the skeletal muscles of human body. It works as energy-enhancing supplement for athletes. A high-intensity, short-duration exercise, such as lifting weights or sprinting, a special chemical reaction takes place to provide a burst of energy to the muscles. Human body does not store very much creatine in the muscles for normal, everyday functions. If it is necessary to increase your muscle strength, improving endurance, and delay fatigue for high-intensity, short-duration sports, then creatine supplements, will be the right choice to achieve the goals.

Antioxidant activity

The antioxidant properties of two raw truffles (*Terfezia claveryi* and *Picoa juniperi*) and five raw mushrooms (*Lepista nuda, Lentinus edodes, Agrocybe cylindracea, Cantharellus lutescens,* and *Hydnum repandum*) have been tested.

Important therapeutic uses of mushrooms

	Species of medicinal mushrooms					
Ailments	*Cordyceps sinensis*	*Lentinula edodes*	*Ganoderma lucidium*	*Grifola frondosa*	*Tremella fuciformis*	*Poria cocos*
Anti-Viral	+	+	+	-	-	+
Anti-Tumour	+	+	+	+	+	+
Anti-inflammatory	-	-	+	-	+	-
Asthma / Bronchial	+	-	+	-	+	-
Blood Pressure	+	+	+	+	+	-
Cardio-Vascular	+	+	+	-	+	-
Chitin	+	+	+	+	+	+
Diabetes	-	-	-	+	+	-
Immune Enhancer	+	+	+	+	+	+
Increase Libido	+	+	-	-	-	-
Kidney Tonic	+	-	+	-	-	-
Lower Cholesterol	+	+	+	-	+	-
Liver / Hepatitis	+	+	+	+	+	+
Stress Reduction	+	-	+	-	-	-

Anti-caries effect

The caries-inhibiting effect of the extract from shiitake (*Lentinus edodes*), has been established. The extract has an inhibitory effect on water-insoluble glucan formation from sucrose by crude glucosyltransferases of *Streptococcus mutans* JC-2 and *Streptococcus sobrinus* OMZ-176. This herbal extract will help the patients from caries, and also from medicines containing carcinogenic effects, used for treating caries.

Mushrooms as a source of antitumour and immunomodulating polysaccharides (Wasser, 2002).

Mushrooms represent an unlimited source of polysaccharides with antitumour and immunostimulating properties. These polysaccharides are of different chemical composition, with most belonging to the group of beta-glucans; these have beta 1, 3 linkages in the main chain of the glucan and additional beta 1, 6 branch points that are needed for their antitumour action. High molecular weight glucans appear to be more effective than those of low molecular weight. Chemical modification is often carried out to improve the antitumour activity of polysaccharides and their clinical qualities (mostly water solubility).

Most of the clinical evidence for antitumour activity comes from the commercial polysaccharides lentinan, PSK (krestin), and schizophyllan, but polysaccharides of some other promising medicinal mushroom species also show good results. Their activity is especially beneficial in clinics when used in conjunction with chemotherapy. Mushroom polysaccharides prevent oncogenesis, show direct antitumour activity against various allogeneic and syngeneic tumours, and prevent tumour metastasis. Polysaccharides from mushrooms do not attack cancer cells directly, but produce their antitumour effects by activating different immune responses in the host. The antitumour action of polysaccharides requires an intact T-cell component; their activity is mediated through a thymus-dependent immune mechanism.

The antiproliferative effect of lectin from *A. bisporus*

Lectins or agglutinins are proteins with affinity for specific sugar residues. Peanut agglutinin (PNA) and the lectin from *A. bisporus,* ABL) both bind to the disaccharide galactosyl beta-1, 3-N-acetyl galactosamine alpha-. This is expressed in keratinocytes as an O-linked chain on CD44, a polymorphic membrane glycoprotein. Many lectins are mitogens and PNA is a mitogen for colonic epithelial cells. However, ABL reversibly inhibits proliferation of colonic cancer cell lines without cytotoxicity and thus has therapeutic potential in situations such as psoriasis where proliferation is increased. The effect of ABL on the growth of normal human cultured keratinocytes and a human papilloma virus (HPV)-transformed cell line has been investigated in detail by **Parslew and co workers (1999)**.

Chapter 18

MUSHROOMS AND RELIGION

The study of psychoactive mushrooms and their uses in different societies began with R. Gordon Wasson's studies. He credited for the coining of term "ethnomycology", the branch of mycology that studies the uses of fungi in various cultures as well as the legends associated with them.

The historical records reveals that mushrooms have been used for less than benign purposes. Claudius II and Pope Clement VII were both killed by enemies who poisoned them with deadly Amanitas. Buddha died, according to legend, from a mushroom that grew underground. Buddha was given the mushroom by a peasant who believed it to be a delicacy. In ancient verse, that mushroom was linked to the phrase "pig's foot" but has never been identified. (Although truffles grow underground and pigs are used to find them, no deadly poisonous species are known.)

Psychoactive mushrooms and their association with various religions was virtually unknown to the general public until the publication of an article entitled "Seeking the Magic Mushroom", in Life Magazine in 1957, by the Maztec Indians of Oaxaca, Mexico.

Wasson was an investment banker by trade, he would go on to study the use of mushrooms in a number of cultures, including those in Mesoamerica, Russia, England, and India. Wasson was a true amateur in every sense of the word related to mushroom and religious cultures. As a banker, he had a little knowledge on mushrooms in the beginning. He developed relationships with linguists, mycologists, chemists, etc. and learned much about ethnomycology.

During his studies, he collaborated with notable personalities such as the French mycologist, Roger Heim, the American ethnobotanist, Richard Schultes and the Swiss chemist, Albert Hofman (the discoverer of LSD). He published several papers on religion and mushrooms including "Soma, Divine Mushroom of Immortality".

It was through his effort that scholarly interest in ethnomycology has reached its present state of study. A great deal of his work was completed following the premature death of his wife, Valentina Pavlovna, in 1958, due to cancer. Wasson decided to retire from his profession as a banker after Valentina passed away, and he pursued his interest in ethnomycology until his death on Christmas Day in 1986.

Wasson and Ethnomycology

Wasson's interest in mushroom can be traced back to a specific incident, in 1927, a year after he was married, when he and his wife, Russian born Valentina Pavlovna were on a delayed honeymoon trip to the Catskills Mountains, in New York. While hiking, Valentina came upon a large cluster of mushrooms that she immediately recognizes as ones that she had often collected in Russia. Her husband became very distressed as he watched her gathering the mushrooms and became more so when she announced that she would cook them as part of their dinner for that night. That evening as Valentina was cooking dinner, Gordon Wasson pleaded with her not to use the mushrooms and said that he would not eat dinner if mushrooms were the part of the meal. Paying no attention to her husband, Valentina cooked the mushrooms for dinner, and as she began to eat them, Gordon Wasson begged her not to eat them and was certain that when he woke the next day he would be a widower. His wife, however, woke in the morning with no ill effects from consuming the mushrooms; they found it interesting how different their attitudes were towards the mushrooms that were found while hiking.

Upon returning from their honeymoon, husband and wife asked their acquaintances, from various cultures, what their feelings about mushrooms were. It was of great interest to them that they found that their acquaintances could be divided into two groups, with respect to their feelings towards mushrooms, mycophobes, those who had a dread and fear of mushrooms and mycophiles, those who had a great love of eating mushrooms. There was no middle ground between these two groups. This cultural difference led them on a quest in documenting the different attitudes, concerning mushrooms and their relationship to culture.

Religious ceremonies and Mushrooms

The beginning of religions goes back to prehistoric time. Regardless to how or when religion came about, one theory concerning religion is now known that in its infancy, religious ceremonies were carried out with hallucinogens or a more popular term, **entheogens**. The use of entheogens, in religion, came about when early man was still-hunting and foraging for food. Most food was utilized to nourish the body and a few were discovered that might have had medicinal use. A few, however, were also discovered that could alter the consciousness of the mind.

The sources of most of the hallucinogens or entheogens were mushrooms that were even believed to allow communication between man and the spirit world where the gods resided. In some cultures the entheogens were even considered to be the gods and as such, these entheogens were considered sacred.

One of the first example of Wasson's work in this area involved what was called Soma in the Rig Vedas, the Sanskrit hymns and songs that became the foundation for Hinduism. Soma was assumed to be a plant for many years until Wasson' work confirming relation of soma with mushroom.

Soma

Wasson provided an exhaustive account of his findings through his book entitled "Soma: Divine Mushroom of Immortality (1967)", tracing all descriptions of Soma, and concluded that Soma was a mushroom and specifically, *Amanita muscara,* the fly agaric.

According to Wasson, Soma, the narcotic god of ancient India was believed to have originated from the Aryans, who invaded India, some 3,500 years back, from the north, which is now Afghanistan, into the Indus Valley, and with them, brought their cult of Soma. Soma is one of the few entheogens that is looked upon as a God rather than a sacred mediator. **Of the 1000 holy hymns in the Rig Veda, 120 are devoted entirely to Soma.**

Amanita muscaria as the Plant of God Soma in the Rigveda

A number of theories for identification of Soma had been circulating for years, but all were found to lack similarities to the poetic descriptions of the plant described in the Rigveda. Some plants including extracts offered up as being Soma, included Bhang (*Cannabis*), Rhubarb, *Periploca aphylla, Sarcostemma brevistigma, Peganum harmala* and *Ephedra vulgaris*. Some of the strongest evidence suggesting these could not be Soma is that the Rigveda makes no mention of the divine plant having any roots, leaves, blossoms, or seeds.

Traces of Amanita in the Rigveda

The Rigveda repeatedly states that Soma grows high in the mountains and nowhere else. For example, Mandala V 43 states that Soma is a "plant from the mountains...," and Mandala IX 46 says that Soma is "seated on the mountain top..."

Amanita grows best in higher elevations surrounding the northern portions of the Indian peninsula, specifically the Hindu Kush range and the Himalayan Mountains. Moreover, Amanita grows as mycorrhizal relationship with a number of different trees, specifically the pines, firs, and above all, the birches, which are mostly found in the high mountains. There is an agreement that the great importance of Soma in Vedic culture is due to the Aryan, who amanita with their migration.

Understanding the description of Amanita in Rigveda through Mandala

An indication that Soma might be the red Amanita is that the Soma is described as being like the sun. "He (Soma) has clothed himself with the fire-bursts of the sun..." (Mandala IX 17). Even before it has fully opened the Amanita can remind one easily of the sun, its white spots seen as the orbs rays. Mandala IX 86 states that "Soma envelopes himself all around with the rays of the sun..." Soma has even been compared to the brilliance of lightning (Mandala IX 22), causing the poet to declare, "make me burn like fire started by friction" (Mandala VIII 48).

A common metaphor for Soma is the bull, the Rigveda's symbol of strength. "He (Soma) bellows, terrifying bull... the hide is of bull, the dress of sheep" (Mandala IX 70). In Mandala IX 97 we find a reference to the "red bull." Might we assume that the skin of the mushroom represents the red hide of the bull, while the woolly spots represent the "dress of sheep"? Many references to a cow's udder can also be found in the Rigveda. Since Soma was "milked" of its juice it seems possible that the Amanita itself could be this udder. This becomes all the more plausible if one has seen an immature Amanita. No doubt to a poetically inspired mind the Amanita could suggest a cows udder. Even the stalk, or *amsu*, is likened to a teat. "When the swollen amsu were milked like cows with full udders..." (Mandala VIII 9). What is also interesting is that the "milk" of Soma is described as a tawny yellow color, a color much closer to cows milk, and to the expressed juice of an Amanita, than would be the green juice of a chlorophyll producing plant.

Another interesting reference is to Soma's being a single eye. "Quickened by the seven minds, he (Soma) has encouraged the rivers free of grief, which have strengthened his single eye" (Mandala IX 9). And in Mandala IX 97, "Soma who has for eye the sun." Though further evidence suggesting that the Amanita is the single eye is lacking Wasson simply asks us to examine an Amanita and inquire if a creeping vine or Rhubarb could fulfill the concept of the single eye so fittingly as the red capped Amanita. Another Mandala about the eye helps bring together the many metaphors descriptive of Soma. "I have drunk the navel (Soma) into the navel (stomach) for our sake. Indeed, the eye is altogether with the sun. I have milked the child of the wise" (Mandala IX 10). In this quote we can see the many metaphors at play, the navel, a word with an archaic history in many cultures and which often has the dual meaning of a mushroom, the eye, which can signify both its shape and its power to open up ones sight, the sun, of which the red cap could represent, and the milking, which is what was done with Soma and which fits into the udder concept of Soma.

The most interesting, and probably the strongest supporting evidence that *Amanita muscaria* is the divine Soma is the mentioning in the Rigveda of there being two forms of Soma. "With those two forms which stand facing us, O Soma, thou reignest over all things" (Mandala IX 66).

Characteristics of *Amanita muscaria*

The *A. muscaria* itself is a bright red mushroom that has woolly white spots on its top that are fragments of the veil from which the mushroom emerges as it explodes out of the ground. These wool like spots, which resemble warts, lent to this mushrooms designation as a "toadstool." The Amanita can grow up to 20 cm tall and nearly 25 cm in diameter once it has fully opened its parasol.

Hallucination from Soma

The light forms of intoxication are accompanied by a certain degree of animation and some spontaneity of movements. Under strong intoxication the senses become deranged; surrounding objects appear either very large or very small, hallucinations set in, spontaneous movements and convulsions. The person intoxicated by the Fly-Agaric sits quietly rocking from side to side, even taking part in conversations with his family. Suddenly his eyes dilate, he begins to gesticulate convulsively, converses with persons whom he imagines he sees, sing, and dances.

Urine drinking as it contains hallucinogens

In Siberia, a ritualistic practice of urine drinking had developed because the tribesmen learned that the psychoactive toxins involved (from *A. muscaria*) passed through the metabolic system unaltered. This practice continued for years together drinking the urine of someone who had consumed the mushroom or who had drunk the urine of someone who had consumed the mushroom. In this matter, the ecstasy of few mushrooms could be extended for several days. An early account of urine drinking, among the Koryak tribe of Siberia was described by Georg Langsdorf, in 1809. The practice of drinking urine was not restricted to Siberia and was also referred to in the Rig-Veda.

Chapter 19

ECOSYSTEM AND MUSHROOM CULTIVATION

Mushroom cultivation is not just a rapidly expanding agribusiness; it's also a significant tool for the restoration, replenishment and remediation of Earth's overburdened ecosphere.

The processing of organic materials into composted soil through mushroom cultivation is a superior use of resources when compared to composting alone. Mushroom cultivation offers many benefits to farmers, food processors, breweries, forest based industries and any one whose operation result in the generation of organic waste. Moreover, it helps to check the depletion of resources, loss of habitat and release of toxic substances into the environment.

Waves of mycelial networks intersect and permeate through one another. The vast, interconnected mantle of mycelia reacts quickly to the availability of plant and animal debris, recycling carbon, nitrogen, sulfur, iron and other essential elements. When storms, floods, volcanoes, or other natural disasters wreak havoc on the environment, fungal champions come to the rescue, capturing debris with mycelium and beginning to recycle. Human beings have the unfortunate distinction of creating more debris than any other organism on the earth.

Saprophytic and parasitic fungi help create the organic components of topsoil, in alliance with myriad numbers of bacteria, insects, and other organisms. An orchestra of primary, secondary, and tertiary saprophytic fungi render wood into biodynamic soil components. These soils benefit plants that in turn use photosynthesis to manufacture their own foods.

Fungi can selectively raise or lower soil pH, increasing the usability of existing soils without the need for additional adjustment prior to use.

In the course of breaking down forest and agriculture wastes, a directed biological succession begins. Composting without generating mushrooms is a lost opportunity for production of food and other valuable products and services.

Nearly all plants have joined with saprophytic and mycorrhizal fungi in symbiosis. Mycorrhizal fungi surround and penetrate the roots of grasses, shrubs and trees, expanding the absorption zone by 10 to 100 fold, aiding in their quest for water and increasing the moisture-holding capacity of soils. This close alliance also forestalls blights and is essential for longevity of the forest ecosystem.

The most progressive communities encourage composting for its refuse, a good first step, but this is far short of the potential that could be realized by making the public mycologically astute.

The first step in helping Nature is to consciously use fungi, mushrooms in particular, in recycling and in the generation of foods. This elementary step was first industrialized by the mushroom growing industry. They have realized that they have taken a revolutionary step in harnessing one of the most powerful biological forces of nature.

Myco-remediation

From a piece of tissue the size of one tenth of your little fingernail, what we call a *clone,* cells can be grown exponentially into millions of pounds of mushrooms in as little as several months. More than 10% of the growing medium or "substrate" (straw, sawdust, compost, most agricultural and forest debris) can be converted into a protein and vitamin-rich food.

Mushrooms are not only the nutritious; rather they have demonstrated abilities in enhancing the human immune system, as they produce natural antibiotics. Yet, it is the residual mycelium in that substrate that holds the greatest potential for ecological rehabilitation.

Mycelia can serve as unparalleled biological filters. Mycelium acts as bleach for saltwater, hence the term "mycofiltration", is used for fungal mats as biological filters.

Mycelium produces extracellular enzymes and acids that break down recalcitrant molecules such as lignin and cellulose, the two primary components of woody plants. Lignin peroxidases dismantle the long chains of hydrogen and carbon, converting wood into simpler forms, on the path to decomposition. By circumstance, these same enzymes are superb at breaking apart hydrocarbons, the base structure common to oils, petroleum products, pesticides, PCBs, and many other pollutants.

After several years, and redundant experiments to prove to naysayers that our data was valid, we have made some astonishing discoveries. (I am continually bemused that humans "discover" what nature has known all along.)

The scientists have discovered that a strain of Oyster mushrooms could break down heavy oil. Analyses showed that more than 95% of many of the PAH (polycyclic aromatic hydrocarbons) have been destroyed, reducing in to non-toxic components, and the mushrooms were also free of any petroleum products.

The mushroom mycelium harbours on several bacterial and killing to clean the environment. The mycelium of mushroom consumes *Escherichia coli* (bacterium found in the sewage water, responsible for several stomach disorders in the human beings), effectively eliminating them from the environment.

Chapter 20

MUSHROOM TOXINS

There are several toxins including Amanitin, Gyromitrin, Orellanine, Muscarine, Ibotenic Acid, Muscimol, Psilocybin, Coprine, which cause Mushroom Poisoning, in human beings.

Mushroom poisoning is caused by the consumption of raw or cooked fruiting bodies (mushrooms, toadstools) of a number of mushroom species. The term toadstool, has been derived from the German word *Todesstuhl* meaning death's stool, is commonly given to poisonous mushrooms.

The toxins involved in mushroom poisoning are produced naturally by the mushrooms themselves. Most mushrooms that cause poisoning in human beings cannot be made nontoxic either by cooking, canning, freezing, or any other means of processing. Thus, the only way to avoid poisoning is to avoid consumption of the toxic species.

Mushroom Poisoning and Nature of Threat

Time of Onset	Type of Poisoning	Nature of Threat
Delayed (6-72 h)	Protoplasmic	Life threatening
Rapid (15 min-2 h)	Gastrointestinal Irritants	Not life threatening
	Neurological	Not life threatening
	Disulfiram-like	Not life threatening

The last three toxins are generally not life threatening, but patient should be provided with appropriate therapy support.

Nature of Poisoning

Mushroom poisonings are generally acute and are manifested by a variety of symptoms and prognoses, depending on the amount and species consumed.

Poisoning of mushrooms is generally categorized by the physiological effects they cause. There are four categories of mushroom toxins:

1. **Protoplasmic poisons:** The poisons that result in generalized destruction of cells, followed by organ failure.

i) **Amatoxins**: It a cyclic octapeptide, and the poisoning by the amanitins, is characterized by a long latent period (range 6-48 hours, average 6-15 hours) during which the patient shows no symptoms. Symptoms appear at the end of the latent period in the form of sudden, severe seizures of abdominal pain, persistent vomiting and watery diarrhea, extreme thirst, and lack of urine production. It is found in several mushroom species, including the *Amanita phalloides, A. virosa, A. verna, Galerina autumnalis*, etc.

ii) **Hydrazines:** Poisoning by hydrazines resembles Amanita poisoning, and it is less in severity. There is generally a latent period of 6-10 hours after ingestion, followed by sudden onset of abdominal discomfort (a feeling of fullness), severe headache, vomiting, and sometimes diarrhea. The toxin is presumed to be related to gyromitrin but has not yet been identified. *Gyromitra esculenta* and *G. gigas* contain gyromitrin, a volatile hydrazine derivative. However, an identical toxin has been reported to be produced by *Verpa bohemica*.

iii) **Orellanine:** The final type of protoplasmic poisoning is caused by *Cortinarius orellanus*. The poisoning produced by this mushroom cause extreme long latent period of 3 to 14 days. An intense, burning thirst (polydipsia) and excessive urination (polyuria) are the first symptoms, which sometimes followed by nausea, headache, muscular pains, chills, spasms, and loss of consciousness.

2. **Neurotoxins:** The compounds that cause neurological symptoms such as profuse sweating, coma, convulsions, hallucinations, excitement, depression, spastic colon, etc., may be divided into three groups, based on the type of symptoms produced.

i) **Muscarine:** Muscarine poisoning is characterized by increased salivation, perspiration, and lacrimation within 15 to 30 minutes after ingestion of the mushroom. Ingestion of *Inocybe geophylla, Clitocybe dealbata* results in an illness characterized primarily by profuse sweating. With large doses, these symptoms may be followed by abdominal pain, severe nausea, diarrhea, blurred vision, and labored breathing.

ii) **Ibotenic acid and Muscimol:** Both the toxins are produced by *Amanita muscaria* and *A. pantherina* mushrooms. Muscimol is approximately 5 times more potent than ibotenic acid. Symptoms of poisoning generally occur within 1 - 2 hours after ingestion of the mushrooms. An initial abdominal discomfort may be present or absent, but the chief symptoms are drowsiness and dizziness.

iii) **Psilocybin:** A syndrome similar to alcohol intoxication is caused by the ingestion of *Psilocybe, Panaeolus, Copelandia, Gymnopilus, Conocybe,* and *Pluteus*. Poisonings by these mushrooms are rarely fatal in adults and may be distinguished from ibotenic acid poisoning by the absence of drowsiness or coma.

3. **Gastrointestinal irritants**: Compounds that produce rapid, transient nausea, vomiting, abdominal cramping, and diarrhea. Numerous mushrooms, *Chlorophyllum molybdites, Entoloma lividum, Tigertop* (*Tricholoma* pardinum), *Omphalotus illudens, Paxillus involutus, Russula emetica, Verpa bohemica*), *Agaricus arvensis* and *Boletus piperatus,* contain such toxins causing gastrointestinal distress.

4. **Disulfuram-like toxins**: Such type of poisoning are generally nontoxic in nature, produce no symptoms unless alcohol is

consumed within 72 hours of eating mushrooms. If liquor has been taken within said period it becomes acute toxic syndrome. The Inky Cap Mushroom (*Coprinus atrimentarius*) is considered both edible and delicious, but if alcohol has been taken after eating, then the coprine present in the mushroom becomes converted into cyclopropanone hydrate, causing fatal effects.

5. **Miscellaneous Poisonings:** Young fruiting bodies of *Laetiporus sulphureus* are considered edible, however, ingestion causes digestive upset, hallucinations in adults and ataxia in children.

Diagnosis of Human Illness

A clinical testing procedure is currently available only for the most serious types of mushroom toxins, the amanitins. The commercially available method uses a 3H-radioimmunoassay (RIA) test kit and can detect sub-nanogram levels of toxin in urine and plasma. Unfortunately, it requires a 2-hour incubation period, and this is an excruciating delay in a type of poisoning which the clinician generally does not see until a day or two has passed.

The exact chemical natures of most of the toxins that produce milder symptoms are unknown. Chromatographic techniques (TLC, GLC, HPLC) exist for the amanitins, orellanine, muscimol/ ibotenic acid, psilocybin, muscarine, and the gyromitrins. The most reliable means of diagnosing a mushroom poisoning remains botanical identification of the fungus that was eaten. An accurate pre-ingestion determination of species will also prevent accidental poisoning in 100% of cases. Accurate post-ingestion analyses for specific toxins when no botanical identification is possible may be essential only in cases of suspected poisoning by the deadly Amanitas, since prompt and aggressive therapy (including lavage, activated charcoal, and plasmapheresis) can greatly reduce the mortality rate.

Diagnoses of Mushroom Poisoning

Symptoms	Cause	Prognosis
Onset Rapid (15 min-2 h after ingestion)		
Nausea, abdominal discomfort and diarrhea, vomiting	Unknown toxins from numerous genera	Rapid and complete recovery; serious cases may last for 2-3 days and require fluid replacement.
Excessive sweating, lacrimation, salivation beginning 15-30 minutes after ingestion	Muscarine from *Clitocybe* or *Inocybe* spp.	Complete recovery within approximately 2 h
Hallucinations without drowsiness or sleep	Psilocybin from *Psilocybe, Paneolus, Gymnopilus, Conocybe*, or *Pluteus* spp.	Spontaneous recovery within 5-10 hrs. if heavy doses it may last for 24 hrs.
Delirium with sleepiness or coma developing within 1 or 2 hrs. after ingestion.	ibotenic acid/muscimol from Amanita muscaria or A. pantherina	Alternating periods of drowsiness and excitement for several h, followed by total recovery
Onset Delayed (6 h-3 days after ingestion)		
Feeling of abdominal fullness and severe headache about 6 h after ingestion, vomiting, no diarrhea	Gyromitrin and related hydrazones from Gyromitra esculenta and its relatives	Complete recovery within 2-6 days; require correction of metabolic acidosis; some deaths have occurred due to liver failure

Persistent and violent vomiting, abdominal pain, profuse, watery diarrhea beginning around 12 h after ingestion	a, b, g, -Amanitins from *Amanita phalloides*, *Galerina autumnalis*, *Lepiota josserandii* and their relatives.	Apparent recovery a few hours after onset of symptoms, followed by a symptom-free period of 3-5 days, which precedes a period of jaundice, loss of strength, coma, and often death
Intense, burning thirst and frequent urination beginning 3-14 days after ingestion, followed by gastrointestinal disturbances, headache, pain in the limbs, spasms, and loss of consciousness	Orellanine from Cortinarius orellanus	Recovery (including recovery of renal function) may require several months in less severe cases; death from kidney failure may occur in severe cases
Onset Conditional (within 72 h of ingestion)		
Flushing, palpitations, rapid heartbeat, rapid, labored breathing occur within 1/2 to 2 h after consuming alcohol, if alcohol was consumed within 72 h of mushroom ingestion	Coprine in *Coprinus atramentarius*	Recovery is spontaneous and complete within a few to several hours after onset of symptoms

Chapter 21
MUSHROOM RECIPES

Mushrooms are very prone to water, therefore while cleaning minimum contact of water should be allowed. Simply wiping them with a dry paper towel or a damp sponge or cloth, or using a soft brush, serves the purpose. If the mushrooms have a lot of soil adhering to the basal portion, trim off the bottoms before use in the culinary. However, caps of most of the mushrooms are edible, cutting them off will loss the weight of mushroom.

Dried mushrooms need to be reconstituted before using them in a recipe. Cover them in a small amount of very hot water and wait until softened.

Broiling/grilling: These methods bring out the meaty flavour in the mushrooms. For the purpose, mushrooms are thread on skewers and grill, or, placed caps stem-side down, on a pan and broil, turning them halfway through the cooking time. Brush them lightly with herb-flavored olive oil before broiling or grilling, and again when turning the mushrooms. Broiling completes in five minutes; grilling takes 10 minutes.

Mushroom Manchurian

Ingredients: Fresh Mushroom 200 gm; Ajinomotto 1 pinch; Ginger Garlic paste 2 tsp; Soy Sauce 1 ½ tsp; Vinegar 1 tsp; White pepper powder ½ tsp; Red chilli powder 2 tsp; Egg One; Corn flour 2 tsp; Chopped green chillies 2 tsp; Chopped garlic 2 tsp; Oil To fry; Salt to taste; Spring onions to garnish.

Procedure: Marinate mushrooms with salt, 1 tsp chilli powder, pepper powder, ajinomotto, soy sauce, vinegar and ginger garlic paste. Keep aside for 3-4 hrs. Beat the egg, add the remaining chilli powder, salt, corn flour and mix well. Put the marinated mushrooms in the egg mixture and mix well until all the mushrooms are well coated. Heat the oil in a pan (high flame), and deep-fry the mushrooms. Fry them for one minute. Heat a little oil, add the chopped green chillies and garlic for a minute and add the fried mushrooms and fry for another minute. Garnish with chopped green part of the spring onions. Serve with fried rice or as a teatime snack.

Mushroom Mutter Paneer

Ingredients: 6 tbsp Ghee or Vegetable oil; 250 gms Paneer cubes; 1 finely chopped Onion; Few chopped Mint leaves; Few chopped Coriander leaves; 3 chopped Green chillies; 3 cloves of Garlic; 1 piece of sliced Ginger; 1 tsp Turmeric, Chilli powder, Garam masala; Salt to taste; 250 gms tiny Button Mushrooms; 250 gms frozen Peas thawed and drained; 175 gms plain Yoghurt mixed with 1 tsp corn flour; Tomatoes and Coriander leaves to garnish.

Procedure: Heat the ghee or oil in a pan and fry the paneer cubes until they are golden brown. Remove and drain on kitchen paper. Grind onion, mint, coriander, chillies, garlic and ginger to a smooth paste. Remove excess ghee or oil from the pan leaving about 1 tbsp. Heat and fry the paste until the raw onion smell disappears and the oil separates. Add mushrooms, peas and paneer. Mix well. Cool the mixture and gradually fold in the yoghurt. Simmer for about 10 minutes. Garnish with tomatoes and coriander and serve hot.

Mushroom Masala

Ingredients: Mushrooms 500 g; Onions 2 nos; Whole red chillies 8-10; Garlic 2 flakes (chopped); Pepper powder ½ tsp; Turmeric powder ¼ tsp; Cumin seeds 1 tsp; Coriander powder ¼ tsp; Green chillies 2 (slit); Coriander Leaves 1 tbsp (chopped);Oil 2 tbsp; Salt to taste.

Procedure: Wash, clean and slice the Mushrooms. Slice the onions. Keep aside ½ the onions. Grind the remaining Onions, red chillies, garlic, pepper powder and coriander powder to a fine paste. Heat Oil in a Kadai, fry cumin seeds for a minute then add the sliced onions. Fry onions for a minute. Then add the sliced Mushrooms. Fry all the three for a minute or so. Then add the ground paste and mix it well add turmeric powder and salt. Mix all together. Then add ½ cup water and allow it to cook. Once the mushrooms are cooked add slit green chilies. Mix together all very well. Cover and cook the mushrooms until the entire water has been reduced. After water is reduced, cook the mushrooms for 15 minutes on medium flame. Once they are fully cooked, garnish with chopped coriander leaves. Serve hot with Chapattis, Puris, etc

Matar Mushroom

Ingredients: 1 lb Mushrooms; 0.5 lb Matar; 1 tsp Coriander Powder; 1 tsp cumin Powder; 1 tsp chilli Powder; A pinch of turmeric; 1 tsp Garam Masala Powder; 3 Onions; 1 tsp Ginger-Garlic paste; 1 lb Tomatoes; 45 gm blanched Nuts (Almonds, Cashewnuts); 1 cup Curd; Few Coriander leaves

Procedure: Grind onions to a fine paste. Grind the nuts separately and add to the onion paste. Heat oil in a kadai, fry the onion paste till golden brown. Now add ginger-garlic paste and fry for few seconds. Add tomatoes, chilli powder, coriander, cumin powder, turmeric powder and fry till oil leaves the masala. Now add mushrooms and peas and fry for 2 min. Add beaten curd and some water and bring to a boil. Let the whole thing cook till the mushrooms and peas are done. Mix well and serve hot garnished with chopped coriander leaves.

Mushroom Korma

Chop fine: 1 lb button mushrooms; Medium sized onions – 3; Greens peas - ½ cup; wash the mushrooms and chop into small cubes; Chop the onions fine.

Spices to add: Cloves – 6; Cinnamon - 1 rolled stick; Tumeric - 1/2 tsp; Salt to taste.

To grind: Medium sized onion - 1 sliced; Tomatoes - 1 large – sliced; Cardamoms – 2; Ginger - 1″ piece; Garlic - minced - 1 tsp; Coconut - ½ grated; Coriander seeds - 1 ½ tsp; Red chillies – 5; Aniseed - 3 tsp ; Grind these into a fine paste . Keep aside.

Procedure: Take 3 tbsp oil in a pan. Add cloves, cinnamon and let them pop. Add onions and fry them till light brown. Add the mushrooms and cook them for 5 minutes on medium fire. Add the turmeric and stir well. Add the ground paste and keep stirring the mix for a couple of minutes. Add water and then cover and cook on medium flame till raw smell disappears. The korma will be ready when comes to top. Granish with coriander and serve.

Mushrooms on Toast

Ingredients: 1 Cup Sliced Mushrooms; Four Slices of toast; White Sauce.

Procedure: In a frying pan heat 1 tbsp of butter and semi fry the mushrooms. Add white sauce (two mugs of milk add 2 tsp full white flour, salt and lots of black pepper powder and mix well before adding to pan) and keep on stirring till thick and then pour over the toast. Serve immediately.

Mushroom Pakoras

Ingredients: ½ Mushroom sliced; ¼ Cup vinegar; ¼ Cup fresh lemon juice; 1 tsp Salt; ½ tsp Cracked black pepper; 1 Cup Besan; ½ tsp turmeric powder; 1 tsp Coriander power 1 tsp; Red chili powder 1 tsp; salt; Oil to fry.

Procedure: Wash the mushrooms in a bowl; add vinegar, lemon juice, salt and black pepper. Add mushrooms to it and stir the mixture well and leave it for 20 minutes. In another bowl take Besan, add salt, turmeric, coriander and chili powder and mix it properly. Add 1 cup of water and make thin pasty mixture. Heat oil in a Pan. Take out the mushrooms from the marinade and add to the second bowl containing besan mixture. One by one soak the mushroom strips in besan and fry until golden brown. Serve hot with ketchup.

Cheese Mushroom Biryani

Ingredients: 1 cup Basmati Rice; 100 g Mushrooms; 150 g Paneer; 1 Onion-sliced finely; 1 Carrot; 1 Capsicum; 2-3 cloves; 1 stick cinnamon; 1 ½ tsp Salt; 2 tbsp Desi Ghee or 3 tbsp Oil.

Procedure: Clean and wash rice. Soak for one hour. Wash mushrooms and cut each into thick slices lengthwise, into 'T' shaped pieces. Make cubes of paneer and fry to a golden colour. Shred the carrot and capsicum into thin strips. Heat ghee. Add cinnamon, cloves and onions. Cook till the onions turn light brown in colour. Add mushrooms, cook for 3-4 minutes on low flame. Add 2 cups of water. Add the strained rice, capsicum and carrot. Add paneer, and salt. Cook covered on slow fire till the rice is done.

Mushrooms Broccoli vegetable

Ingredients: Mushrooms 300 g; Broccoli 1 bunch; Onions 1 big; Garlic 5-6 flakes; Tomatoes 1 big; turmeric ¼ tsp; Cumin powder ½ tsp; coriander powder ¾ tsp; Garam masala ½ tsp; Oil 2 tbsp; Salt to taste.

Procedure: Cut mushrooms into halves. Separate small florets of broccoli. Chop onions, garlic, and tomatoes into fine bits. Heat oil in a frying pan; add garlic, onions and fry till crisp. Add mushrooms and fry till the water starts leaving. Continue frying till half of the water simmers. Now add broccoli and fry for few second. Sprinkle all the dry spices except garam masala and let the vegetables cook. Add tomatoes and fry till tomatoes get softer. Sprinkle garam masala and remove from fire. Serve hot with Puris or Chapattis.

Mushroom Rice

Ingredients: 1 Big Onion; 10 Garlic flakes; 1" Ginger piece; 7 Green Chillies chopped; Mint Leaves a palmful chopped; Coriander Leaves chopped.

Procedure: Heat oil in a pan. Add onions chopped and the rest of the items. Fry mushrooms well in separate pan. Mix both the fried materials and to cooker cooking for rice. Add turmeric powder separately. Close the vessel and mix well for 5 minutes. Garnish with coriander leaves and grated cheese.

Mushroom Curry

Ingredients: 100 g Mushrooms; 2 Onions; 2 Tomatoes; 5- sticks Cinnamon; 3-4 Cloves; Ginger Paste; Green peas; Red Chili Powder (2 tsp); Salt to taste.

Procedure: Heat oil in a pan. Add Onions and cinnamon and fry for a while. Then add chopped tomatoes, ginger, garlic paste and fry for a while. Add Red Chilli powder, salt and sprinkle some water. Add the chopped mushrooms and stir it well. Close the lid and let the mushroom cook. Add Green peas and let it cook along with the mushroom. Keep the lid closed for a while. Once Mushroom cooked, garnish with coriander leaves and Serve with Chapattis.

Mushroom Paneer gravy

Ingredients: Mushroom 100 g; Paneer 100 g; Big onions 4, chopped; Tomato 2, chopped; Salt, Chilli powder to taste; Turmeric powder, a pinch; Cumin Powder 1 tsp; Garam Masala, a pinch; Butter 2 tsp; Oil 2 tsp; 50 g cashew nut ground to a paste; Coconut milk one and a half tumbler.

Procedure: Heat oil and butter in a pressure cooker. Add onions and fry till light brown. Add tomatoes and fry a little. Add Paneer, Mushroom, Coconut milk, cashew paste and all the dry powders.

Keep in high flame and after a whistle keep in slow flame for 15 minutes. The aroma of the gravy will start coming from 10 minutes onwards.

Capsicum Mushroom Fry

Ingredients: Capsicum 3; mushroom 100 g; onion 1; Green chillies 2; Ginger little; Chillie powder ¼ tsp; Tomato sauce 3 tbsp; Oil 1 tbsp; salt to taste.

Procedure: Cut onions finely, capsicum lengthwise and slice mushroom. Heat oil in a pan, fry onion; add green chillies and ginger. Add capsicum pieces and fry. Add mushroom, chillie powder, salt and fry for few minutes. Finally add tomato sauce and fry. Serve hot.

Mushroom Rolls

Ingredients: Bread 1 loaf; Mushrooms 100 g, chopped; Onions - 1 big, chopped; Green chillies - 4 or 5, minced; Cumin seeds - 1 tsp; Salt to taste; Turmeric powder, a pinch; Oil to fry.

Procedure: Heat oil in a pan; add the cumin seeds, chillies and the onions. Fry the onions till they are translucent and then add mushrooms. Add salt and turmeric and keep frying till the mushrooms are cooked. Meanwhile, take water in a shallow plate and dip each bread slice and remove quickly, taking care that the bread slices don't get too soggy. Squeeze out the water and arrange the bread slices on a baking sheet coated with cooking spray. Spoon out the filling on each of the slices and roll each slice, sealing the ends with water. Bake the slices in oven for about 10 minutes or till the bread slices turn golden brown. Serve hot with tomato ketchup.

Mushroom Baingan Bharta

Ingredients: 1 large Eggplant; 10 Button Mushrooms; ½ cup

Green Peas; 2 large Onions; 2′ Ginger; 4-6 Garlic flakes; 2 Tomato; Oil; Salt to taste; Garam-Masala and Chilli Powder.

Procedure: Roast the eggplant in an oven. Peel the skin and churn it in a blender. Fry onions in a pan till golden, add ginger and garlic paste to it. Fry for sometime then add the mushroom and green peas. Add salt, chilli powder, garam masala and tomatoes. Cook for 4 minutes and then add the roasted eggplant. Cook for some more time. Serve hot with Rotis or Nans or Paratha.

Fried Mushroom

Ingredients: White Mushroom 100 g; Onion 1; Tomato 1; Chopped ginger; Cumin seeds; Coriander powder ½ tsp; Red chillies powder 1 tsp; Turmeric 1 pinch; Chopped green chillies 3-4; Chopped coriander leaves; Garam masala powder; Oil; Salt to taste.

Procedure: Clean and cut the mushroom into small pieces and keep it aside. Heat oil in pan. Add cumin seeds, when starts cracking add onion and fry until light brown. Add ginger, green chillies, coriander powder, turmeric powder and cook for one minute. Add tomatoes, red chillies powder and cook for 5 minutes. Add Mushroom and cook for another 10 minutes. Add garam masala powder and let lid open for next 5 minutes. Garnish with coriander leaves. It is ready to serve with rice.

Mushroom Sandwich

Ingredients: Button mushrooms – 250 g; Green peas – 50 g; Onion – 1; Garam masala – ½ tsp.; Chilli powder – ½ tsp; Sandwich bread – 6-8 slices; Butter – 1 tbsp; Oil- 1 tbsp; Salt to taste.

Procedure: Chop onions and mushroom. Boil green peas. Fry onion in pan for a minute. Add mushrooms, green peas and cook for sometime. Add salt, garam masala, chilli powder and keep stirring till cooked. Keep it aside. Take bread slices, stuff the filling, and toast in a sandwich maker. Now sandwich is ready.

Malai Mushroom Curry

Ingredients: Button Mushrooms 200 g; Onion 1 large; tomato 1 large; Ginger ½" grated; Garlic 2 flakes; Turmeric powder ½ tsp; Red chilli powder ½ tsp; Coriander powder 1 tsp; Cumin powder ½ tsp; Mango powder ½ tsp; cream 3 tbsp; Fresh coriander to garnish.

Procedure: Chop the onions and tomatoes, cut the button mushrooms in halves. Take 2 tbsp of oil in a pan and heat it. Add garlic to oil followed by onions and ginger to it. Fry till onions turn a light brown, add tomatoes and fry for a while till tomatoes are slightly cooked and mixed. Add mushrooms to it. Cover and cook till mushrooms are half cooked. Finally add cream to the mushrooms and cook till the mushrooms absorb the cream. Garnish with chopped coriander. It is ready to serve.

Mushroom Cashew Walnut Patties

Ingredients: 118 ml coarsely ground raw walnuts; 118 ml coarsely ground raw cashews; 480 ml cooked short grain or sweet brown rice; 240 g button mushrooms, finely chopped in the food processor; 1 flake garlic; finely minced ¼ tsp pepper; ½ tsp salt; 1 tbsp EnerG Egg Replacer; 4 tbsp water; 2 tbsp organic canola oil.

Procedure: Combine all ingredients except the canola oil together in a large mixing bowl. Mix well. Heat canola oil in a large skillet over high heat until just hot enough for a drop of water to sizzle. Form mushroom mixture into patties and fry for 2 minutes on each side or until crisp. Makes about 12 small patties.

Chapter 22

MAGIC AND HALLUCINOGENIC MUSHROOMS

MAGIC MUSHROOMS

The mushrooms that contain psilocybin are called magic mushrooms. Psilocybin is a psychedelic drug with effects similar to that of LSD (**Lysergic acid diethylamide**). Psilocybin mushrooms have been used by many indigenous cultures to induce altered states of consciousness during religious rituals.

Magic mushrooms are either mixed with food, brewed into a tea, or eaten raw that too either fresh or dried. There is no, so far, any predictable way of estimating the amount of psilocybin in each mushroom. The amount is determined by the strain, size and age of the mushroom species.

Magic mushrooms, at low doses, produce feelings of relaxation, as caused by the use of cannabis. Users often report laughing a lot and finding things funnier than they would normally. At higher doses, the experience is closer to that of LSD, intensifying colours and producing visual hallucinations and feelings of euphoria. The experiences magic mushrooms are psychedelics, increasing emotional awareness and causing less psychological confusion in some users. Many users find the mushroom experience to be spiritually significant while others find it frightening. The effect of magic mushrooms lasts for about four to five hours.

The biggest danger associated with magic mushrooms is misidentification. User is taking magic mushroom, but due to misidentification, he takes some poisonous mushroom, which cause stomach pains, vomiting, diarrhoea and even death.

Magic mushrooms, like all hallucinogens, can trigger underlying mental disorders and cause schizophrenic-type symptoms. If user has taken magic mushrooms, he should avoid driving as hallucinogens present in magic mushroom cause impair judgment.

Magic Mushrooms are illegal. Possession can result in long prison terms. Supplying mushrooms to someone else (whether or not money was exchanged) carries even longer sentences.

HALLUCINOGENIC MUSHROOMS

Some mushrooms contain toxic and/or hallucinogenic compounds known, as hallucinogens. These hallucinogenic substances alter or produce false perceptions of sight, sound, taste, smell or touch upon ingestion. Some toxic substances in mushrooms can cause severe illness and even death. Most hallucinogenic substances, including LSD (Lysergic acid diethylamide), are even illegal.

The hallucinogenic properties of certain mushrooms have been known for centuries. These mushroom have been known to be used by people during religious ceremonies. The Aztecs used the term "teonanacatl" meaning, "flesh of the gods" to describe hallucinogenic mushrooms. Historians have proposed that Aztec spiritual leaders used these hallucinogens to induce an altered state of consciousness that they believed would allow them to communicate with their gods and other spirits.

Lysergic acid diethylamide (LSD)

LSD is an odorless, colorless and tasteless water-soluble powerful drug that alters a user's mood, thoughts or perceptions. It is grouped into a class of drug known as **hallucinogens** or **psychedelics**. These drugs can cause auditory, visual or somatosensory hallucinations, paranoia or dream-like states.

LSD was first synthesized from a fungus (*Claviceps* spp.) that grows on rye and other grains. In 1938, Albert Hofmann working in the Swiss pharmaceutical company called Sandoz, produced LSD for the first time. He was hoping that this new drug could be used to stimulate circulation and respiration. However, the tests he conducted were all failures and he forgot about LSD for 5 years. In 1943, Hofmann accidentally ingested (or somehow absorbed) a bit of LSD and experienced some of the psychedelic effects including dizziness, visual distortions and restlessness, of this chemical. A few days later he prepared 0.25 mg of LSD in water and drank it. He again experienced the mood and could realize the altering effects of LSD.

Some behavioural effects which are noticed upon consumption of LSD include, feelings of strangeness; vivid colors; hallucinations; confusion, panic, psychosis, anxiety; emotional reactions like fear, happiness or sadness; distortion of the senses and of time and space; flashback reactions. In addition to these increases in heart rate; blood pressure; and muscle weakness are also noticed.

Effect of LSD is quite dependent on the dose the user takes. A dose as small as a single grain of salt, about 0.010 mg, can produce some effects. Psychedelic effects are produced at higher doses of about 0.050-0.100 mg. The effects of LSD depend on a user's mood and expectations of what the drug will do and last several hours.

LSD has a chemical structure that is very similar to the neurotransmitter called **serotonin**. It is thought that the effects of LSD are caused by stimulation of serotonin receptors on neurons, perhaps in the brain area called the **raphe nuclei**. However, it is still not clear what produces all the effects of LSD.

Psilocybin and psyilocin

The term "psilocybin" has been derived from the Greek words "psilo" meaning "bald" and "cybe" meaning "head." Both psilocybin and psyilocin produce yawning, inability to concentrate, restlessness, increased heart rate, and hallucinations both visual and auditory. These symptoms may appear in 30 to 60 minutes after the mushroom is eaten and can last about four hours.

The chemical structure of psilocybin and psilocin is similar to the neurotransmitter called serotonin. In fact, the primary effect of psilocin is on the receptors for serotonin. There is also evidence that psilocybin reduces the reuptake of serotonin by neurons in the brain allowing this neurotransmitter more time to act in the synapse.

Muscimol and ibotenic acid

Muscimol and ibotenic acid from the *Amanita muscaria* appear to act on the GABA neurotransmitter system. Muscimol activates GABA receptors on neurons. The GABA neurotransmitter system is one of the brain's major inhibitory systems. Therefore, muscimol acts to inhibit the activity of neurons in the brain.

Muscimol produces feelings of euphoria, hallucinations, muscle jerks, drowsiness, sweating, pupil dilation, and increased body temperature. Symptoms appear in 30 to 90 minutes after eating this mushroom and are most intense after two or three hours. People who eat these mushrooms usually fall into a deep sleep. Some people describe the effects of eating *Amanita muscaria* as similar to being intoxicated by alcohol.

Examples of Hallucinogenic Mushrooms

1. Mushrooms containing Psilocybin/Psilocin

Mushrooms that contain these hallucinogens include include *Psilocybe mexicana, Stropharia cubensis, Psilocybe semilanceata, Psilocybe pelliculosa, Panaeolus subbalteatus, Psilocybe cyanescens, Psilocybe baeocystis.*

2. Mushroom containing muscimol and ibotenic acid

The *Amanita muscaria* is also known as "fly agaric" because of its ability to attract and kill flies. This mushroom species contain hallucinogenic chemicals include muscimol and ibotenic acid.

Other species of genus *Amanita* including *A. virosa, A. verna* and *A. phalloides* called deadly mushrooms and contain toxins that destroy cells in the liver and kidneys.

Chapter 23
EDIBLE AND POISONOUS MUSHROOMS

Much has been discussed about the edibility of mushrooms, being source of human food. More than 2000 species are said to be edible including those mentioned in the following text. There are several mushroom species quite similar to those of edible ones, called poisonous mushrooms. These cannot be easily distinguished without an experience. There are also some mushroom species, which lose their toxic effect while cooking. A Croatian Proverb says *"All mushrooms are edible, but some only once."*

Edibility Rules for mushrooms

There are following rules, which need to be taken care before consuming a mushroom:

1. **Don't eat anything if you aren't sure of its identification**: Before you eat a new species of mushroom must compare it with a confirmed specimen of that species, or get your identification confirmed by an experienced mushroom specialist.
2. **When trying a species for the first time, eat only a small amount:** The safest method for most mushrooms is to take a few small pieces and eat them either alone or with a bit of a staple such as bread or rice. Wait a day before trying more. Like all foods, mushrooms should generally be eaten in moderation.
3. **When trying a species for the first time, always cook it:** Most species of mushrooms are more digestible when cooked or dried.

4. **Never mix two species that you haven't eaten before:** If you have an allergic reaction, you won't know which species has caused it.
5. **Don't eat rotten mushrooms:** Most "mushroom poisoning" cases are actually due to food poisoning, which may be caused by the stale food.
6. **If you** have **any doubt that a mushroom is edible, don't eat it**: If you think you're going to get sick, you probably will convince your body that it should get sick.
7. **Don't experiment:** There is an old saying, "There are old mushroom hunters, and bold mushroom hunters, but there are no old, bold mushroom hunters."
8. **Do not eat mushrooms raw**: Only eat cooked mushrooms.

Myths regarding Edible and Poisonous Mushrooms

There is no general rule for identification of poisonous and non-poisonous mushrooms. There are several traditional methods, which are unreliable, as

1. Traditionally, it was believed that edible mushroom are peeled off easily and did not change a silver spoon black while cooking. This belief went on incorrect as *Amanita phalloides*, which is deadly poisonous, peels off easily and have no effect on silver spoon while cooking.
2. It was also held that creamy ones were edible, which was again wrong as *Amanita phalloides, A. verna* and *A. verosa* are same in colour but are deadly poisonous; however, this may be true for *Cantharellus cibarius* and *Tricholoma nuduns* which are bright coloured and edible.
3. It was thought that feeding doubtful specimens to dogs or cats or rabbits may connote the edibility of mushrooms, but it led to the non-of use as in a particular animal there may be different reaction, e.g. stomach contents of rabbits are able to neutralize the poison of the most of dangerous species of mushrooms.

4. It is not safe to believe that fungi, which have been nibbed by slugs or larvae, are not poisonous. This is also wrong that some slugs are fond of some species of mushrooms e.g. *Amanita phalloides.*
5. Some of the specialist refer to colour and shape, smell and taste, exudation of milk when broken or change in colour of flesh, coagulation of milk, etc. as pointers to the edibility or otherwise of the mushrooms, but these are also incorrect, for instance, the yellow staining mushroom *Agaricus xanthoderma* (highly poisonous mushroom) quickly changes from white to yellow when it is rubbed and later changes from yellow to brown to chocolate colour. Similarly, *Boletus satanas* changes to bluish-green when it is damaged, but it is also poisonous.
6. It was also thought that one genus, mostly having the edible species, would not include the poisonous ones but this held as wrong notion because in one genus some of the species may be edible whereas others may be poisonous, e.g. genus *Agaricus* include both edible (*A. bisporus, A. arvensis, A. campestris*) and poisonous (*A. xanthoderma*). Also genus *Amanita* is said to be related with poisonous mushrooms (*A. phalloides, A. verosa, A. verna*) but it also contains edible species (*A. rubescense*).
7. *Gyromitra esculenta* was sold once in the European and American markets as a common edible mushroom having good taste, after sometimes it was banned as it proved to be highly poisonous mushroom and was responsible for several lives. However, for the same reason some researchers believe that there may be two species of the above mushroom of which one may be edible whereas other may be poisonous. Some other believe that it may contain toxin, which is water soluble and leached to the water used for cooking; if water poured off, the mushroom becomes edible. It is worthwhile to note that some people may be allergic to some of the species of mushroom. Also mushrooms cause illness if taken in the meals and simultaneously one takes alcohol, e.g. eating *Coprinus atramentarius* with alcohol causes purplish red skin discolouration as a temporary result.

True or False about the edible and poisonous mushrooms

- Poisonous mushrooms tarnish a silver spoon. **False**
- If mushroom peels off easily, you can eat it. **False**
- All mushrooms growing on wood are edible. **False**
- Mushrooms that squirrels or other animals east are safe for humans. **False**
- All mushrooms in meadows and pastures are safe to eat. **False**
- All white mushrooms are safe. **False**
- Poisonous mushrooms can be detoxified by parboiling, drying or pickling. **False**

Some Poisonous Mushrooms which look alike Edible Mushrooms

Poisonous species	Appearance	Mistaken for
Amanita tenuifolia	Pure white	*Leucoagaricus naucina*
Amanita bisporigera, A. verna, A. virosa	Pure white	*Amanita vaginata, Leucoagaricus naucina,* white *Agaricus* spp., *Tricholoma resplendens*
Amanita phalloides	Pure white variety	*Amanita citrina, A. vaginata, L. naucina, Agaricus* spp., *T. resplendens*
Buttons of *A. bisporigera,. A. verna, A. virosa*	Pure white	Buttons of white forms of *Agaricus* spp. Puffballs such as *Lycoperdon perlatum*, etc.
Amanita phalloides	Green = normal cap color	*Russula virescens, Amanita calyptrodermia, A. fulva, Tricholoma flavovirens, Tricholoma portentosum*

Amanita phalloides	Yellow variety	*Amanita caesarea*
Amanita brunnescens		*Amanita rubescens A. pantherina*
Galerina autumnalis	Little Brown Mushroom (LBM)	*Gymnopilus spectabilis, Armillaria mellea*
Leucoagaricus brunnea, Lepiota josserandii, L. helveola, L. subincarnata	LBM	*Lepiota* spp., *Leucoagaricus* spp., *Gymnopilus* spp.

IMPORTANT EDIBLE MUSHROOMS

Button mushrooms: These mushrooms are the most common variety prepackaged in supermarkets; available fresh, canned, or frozen. White mushrooms are mildly flavored, are tasty when eaten raw but even more flavorful when cooked.

Chanterelles: These trumpet shaped fungi are highly regarded mushrooms favored for their gold to yellow color, and rich flavor, ranging from apricot to earthier tasting. Chanterelles are best eaten fresh, although they are also available dried or canned.

Crimini: These mushrooms are similar to the button variety, yet they are darker in color, have a richer flavor, and have a more dense texture. Criminis were once an imported mushroom but are now grown domestically.

Morel: These mushrooms are highly priced and highly prized for their intense earthly flavor. They are usually found in the wild, although can now be grown commercially. This conical shaped, honey combed surface fungi is small, with dark brown hues, is suitable for stuffing and is ideal for sauces and stews.

Oyster: These mushrooms grow in clusters, and range in color from off-white to shades of brown. Subtly tasting like an oyster, its chewy texture is more suited to cooked dishes.

Shiitake: These mushrooms were originally cultivated on natural oak logs and only grown in Japan, but are now available domestically. These mushrooms are large, black-brown, and have an earthy rich flavor. This fungus is enjoyed in stir-fries, soups, or even a meat substitute. Dried Shiitakes have more intense flavors and are sometimes preferable to fresh.

Other edible mushrooms

Agaricus arvensis	It is found in lawns, meadows, cultivated fields, and in pastures, etc. either solitary or scattered. Pileus is smooth, white or yellowish, convex or conical, bell shaped.
Agaricus campestris	It is found growing in fields, pastures and meadows. Cap is generally globular in shape. The colour varies from creamy to white to light brown.
Agrocybe praecox	This fungus is found in logs, lawns and characterized by a convex (yellowish) pileus at the beginning turning expanded (yellowish brown) at the maturity. The species if found distributed in western Uttar Pradesh.
Amanita caesaria	The orange coloured mushroom with attractive pileus characterized by smooth, hemispherical, bell shaped and slightly curved from the margins. The annulus is hanging down to the stem looking like collar.
Amanita vaginata	The fungus is varying in colour from white, brown or yellow. It is found in open places having long volva forming a sheath. The pileus is bell shaped, fragile in nature.

Armillaria mellea	It tastes like a good all-purpose mushroom, found growing in dense clusters, in thin woods or open places. The cap is fleshy, and honey coloured (hence honey mushroom).
Bovista plumbea	This mushroom species is quite small found growing nearby Mussoorie area on the ground in old pastures. The fruit body is whitish and glabrous.
Calocybe indica	It grows in nature in Karnataka and Tamil Nadu on humus soil under the roadside trees or in forests.
Cantharellus cibarius	It has a pleasant, fruity-peppery taste and a texture that is a bit more delicate than Agaricus species. The rich egg coloured pileus is fleshy, convex at beginning turning to expanded at maturity. The cap appears like inverted cone.
Coprinus attramentarius	This species found growing as dense clusters in damp rich ground, having egg shaped pileus with grey colour. The cap is smooth but slight scaly appearance in present as margins are ribbed. This species contains class D toxin called corprine and need to eaten with care. The onset of symptoms is related to the consumption of alcohol three hours to as long as 5 days after the meal of mushrooms.
Coprinus comatus	This species is good as an appetizer in soups. The flavor is subtle but distinctive. It is found growing on grassy land, and refuse dumps. The pileus is fleshy, egg shaped at first becoming bell shaped at maturity. The splitting is noticed along the margins.

Coprinus micaceus	The fungus is found growing on ground or sometimes in the bases of living trees. The pileus is oval shaped with yellow to tan colour.
Flammulina velutipes	This mushroom is grown on the standard sawdust recipe and can be grown in a variety of containers, including the filter bag. Commercially, it is grown in jars. Several fruiting bodies emerge from the common rhizoids. It is relatively fast growing species. Grown in clusters, they develop long thin stems, about four inches, with tiny little caps, the largest being the size of a pencil eraser. With their delicate ivory color and dainty appearance, they're prized for their ability to provide a simple yet dramatic garnish.
Flammulina velutipes	It is found growing on dead logs emerging as several fruiting bodies from a common root. The pileus is flattened having glabrous surface and upward rolled margins. The lower half of stipe is covered with dense reddish brown hair giving a velvety appearance, without annulus and volva. It contains a polysaccharide, which is apparently an immune system stimulant that may have indirect anticancer properties. It also contains a red cell-destroying toxin.
Hirneola auricular judae	It was previously called as Auricularia auricula, found growing on dead wood including tree trunks. It looks like ears, which give the name "Jews ear".

Hydnum spandum	The cap is varying in colour from buff, pale brick red to white. It is found growing single or in clusters fleshy cap having convex shape.
Morchella esculenta	It is found growing in apple orchards and open grass lands, having a cylindrical blunt toped cap. The pits are elongated and narrow.
Pleurotus ostreatus	It is a good all-purpose mushroom, found growing in clusters on dead tree trunks.

IMPORTANT POISONOUS MUSHROOMS

Amanita phalloides	It contains a mixture of cyclopeptides known as amanitins in potentially lethal levels. The first symptoms generally appear 6-12 hours after ingestion. It resembles with popular edible Volvariella species.
Amanita ocreata	It causes poisoning which is characterized by sharp abdominal pains, nausea, vomiting and watery diarrhea, if poisoning persists it may cause jaundice, liver and kidney failure, convulsions and coma. Death usually occurs in 4-7 days.
Lepiota clypeolaria	It is a small, poisonous mushroom. The specific toxin is not known. It is recognized by a white spore print, white gills that are not attached to the stipe.
Galerina autumnalis	It contains lethal levels of amatoxins. It is easily mistaken with Armillaria mellea, but A. mellea has white spore prints whereas, Galerina species yield brown spore prints.

Amanita smithiana — The food poisoning of this species causes vomiting, abdominal pain and diarrhea followed much later by reduced ability or inability to urinate.

Gyromitra esculenta — This species contains gyromitrin or related hydrazines. Gyromitrin is water-soluble and it vaporizes on cooking. Victims experience sudden headache, a bloated feeling, nausea, vomiting, stomach pains, watery or bloody diarrhea, and fatigue or dizziness.

Clitocybe clavipes — It causes similar effects to Coprinus atramentarius, if alcohol is consumed within several days of a meal of the mushrooms.

Amanita muscaria var. formosa — It is a color variant of *Amanita muscaria*. It may be white, yellow, red, orange or even brown. It is characterized by white gills, stem, and a white spore print.

Clitocybe dealbata — It contains a high level of muscarine and related compounds.

Hebeloma sinapizans — This poisonous species causes severe gastrointestinal distress. It contains contain unusual sesquiterpines and another contains toxic triterpine glycosides.

Tricholoma pessundatum — This species is reportedly poisonous, causing violent vomiting, diarrhea and weakness lasting up to three days. It is sometimes confused with edible *Tricholoma populinum.*

Tricholoma sulphureum — It resembles to the edible species *Tricholoma flavovirens.* It is characterized by a foul smell of rotten egg and coal tar gas that should discourage anyone from eating them. It contains a hemolysin that could attack red

	blood cells. However, hemolysin can be destroyed in cooking.
Agaricus hondensis	It is poisonous species causes flu-like symptoms when eaten.
Stropharia coronilla	It resembles with *Agaricus campestris,* except for pink free gills. Victim after consumption suffer headache, vomiting, generalized aching, confusion, loss of coordination, dizziness and hallucinations.
Lactarius scrobiculatus	It has white milk that soon turns a distinctive yellow and has a very acrid taste. The acrid species are very irritating to the gastrointestinal tract.

Chapter 24

GLOSSARY

Acrid: Severe burning sensation in the tongue. Acrid is used as indicator for identifying poisonous species.

Adnate: Pertaining to the attachment of the fertile tissue (the gills, tubes, spines, etc.) to the stipe of the fungus in which the attachment is typically perpendicular into the stipe, *i.e.* without dipping towards the pileus or down the stipe.

Adnexed: Pertaining to the attachment of the fertile tissue (the gills, tubes, spines, etc.) to the stipe of the fungus in which the fertile tissue typically curves upwards towards the pileus of the fungus before attaching to the stipe.

Agaric: A term commonly used to describe a fungus having a cap (pileus), gills (lamellae), and a stem (stipe), *i.e.*, what most people would call a mushroom.

Amyloid: A chemical staining reaction in which the tissue, spore wall ornamentation, etc. stains bluish-black in Melzer's reagent. Examples include the spore ornamentation of species in the genera *Russula* and *Lactarius*.

Annulus: A ring of tissue on the stipe of mushroom formed by the rupture of a membrane (the partial veil) connecting the cap and stipe of a developing mushroom. A special layer of tissue that connects the margin of a mushroom pileus to the stipe that can either form a ring around the stipe, hang as fragments from the margin of the pileus, or be variations of the two. Examples may be found in many genera such as *Amanita, Cystoderma, Lepiota,* and *Suillus,* among others.

Apiculus: It is a particular point at the end of spore where it is attached to the basidium.

Apothecium: An open Cup or saucer shaped fruiting body of Discomycetes.

Applanate: Flat structure.

Ascus: A specialized sexual reproductive cell found in the fertile area of the hymenium of all Ascomycetes. An ascus is typically club shaped and which forms internally 4 or 8 ascospores, usually in a row.

Attenuated: Narrowed or tapering.

Basidium (Pl. Basidia): A specialized sexual reproductive cell found in the fertile area of the hymenium of all Basidiomycetes, typically shaped like a baseball bat. A basidium possesses four slightly inwardly curved horns (sterigma) to which the basidiospores are attached.

Biological efficiency (B.E.): It is defined as the ratio of fresh weight of mushrooms to dry weight of compost at spawning expressed as a percentage.

Caeilocystidium: Cystidium developing at the edge of a gill.

Carminophilous granulation: The granules present in some basidia have affinity for aceto-carmine. The granules turn black after treatment with.

Cespitose: Clustered mushrooms fused at the base. Also spelled caespitose.

Cladistics: Deals with the systematic classification of groups of organisms on the basis of the order of their assumed divergence from ancestral species.

Clamp connection: A special connection, which forms at the junction of two adjacent fungal filamentous cells. A clamp connection looks something like the handle on a coffee cup. However, it may be flattened against the wall of the cells or may have a large opening (in this case a keyhole clamp) that allows the migration of nuclei between developing cells.

Cortina: A special type of annulus that is filamentous, resembling a spider web, attached from the margin of the cap (pileus) to the stem (stipe) when young. In age only a few fibers may remain on the cap margin or the stipe.

Cystidium: A specialized sterile end cell formed anywhere in fungal tissue. It is most commonly found in the hymenial layer of tissue, but may also be found on the surface of the cap, the surface of the stipe, or even within the sterile tissue of the stipe. There are many different types of cystidia; they are named based on the location where they are found, *e.g.* Dermatocystidia- on the surface tissues; Pileocystidia- found on the surface of the pilius; Caulocystidia- found on the surface of the stipe; Cheilocystidia- on the edge of the gill; Pleurocystidia- on the face of the gill; Endocystidia- form in the tramal tissue of the cap, or stipe; OR on their morphology, function, chemical reactions etc. such as Leptocystidia-which are thin-walled, smooth and do not have distinctive contents and are not tramal in origin; Gloeocystidia- which are variable in shape and stain easily or have conspicuous contents; Lamprocystidia- which are thick-walled and without conspicuous contents, etc.

Decurrent: Pertaining to the attachment of the gills to the stipe, in which the gills curve partly down the stipe towards the base of the stipe.

Deliquescing: The process by which gills in the genus *Coprinus* rapidly break down into a black ink-like liquid, droplets of which disperse spores.

Dextrinoid: A chemical staining reaction in which the tissue, spore wall ornamentation, etc. stains reddish to reddish-brown in Melzer's reagent.

Divergent — usually referring to gill trama, in which the tramal hyphae branch outward from the gill center towards the hymenium and downward towards the gill edge.

Eccentric: Away from the centre.

Echinulate: Referring to spiny ornamentation; *e.g.* the fine spines seen on the surface of some species of puffballs in the genus *Lycoperdon*. Also, the finely spined spores of *Laccaria* species.

Economic Injury Level (EIL): The number of pests which, if not controlled, will result in economic loss mushroom crop.

Economic Threshold Level (ETL): The minimum number of pests needed before control measures are initiated. The ETL is also called the **Action Level** (AL) and it occurs early in the crop and is the point when preventive chemical controls should be applied.

Evanescent: The structures such as annulus, veil, scales, etc., which develop slightly and disappear rapidly.

Farinaceous: An odor variously described as that of raw potatoes, raw cucumbers, or even of soaps; mealy.

Ferruginous: Rust coloured.

Fibrillose: Possessing surface fibrils.

Filamentous: Composed of thread-like cells.

Floccose: Having a cottony appearance, like a flocked Christmas tree. Seen in some species of *Amanita*.

Globose: Round in shape.

Gregarious: A growth form in which mushrooms fruit in relatively close proximity.

Hygrophanous: Having the characteristic of changing colour upon drying.

Hymenium: A fertile layer that bears either baidia and basidiospores or asci and ascospores, working as supporting part of agarics.

Imbricate: Overlapping one another as the shingles on a roof.

KOH: The chemical Potassium Hydroxide. Used in a 3% solution, it is a standard mounting medium used to rehydrate material for microscopic examination. It may also be used as a macro- or micro-chemical reagent differentially staining the tissues of some species. Concentrated solutions of this chemical are caustic and should be handled with care.

Lamellae (or Gills): The technical term used to describe the gills of a mushroom.

Lignicolous: Living on wood.

LSD (Lysergic acid diethylamide): It is a chemical that alters a user's mood, thoughts or perceptions, hence they grouped into a class of drugs known as **hallucinogens** or **psychedelics**.

Lumber: Wooden inventories of farm or mushroom house not presently in use.

Macrofungi: Fungi visible to the naked eye.

Magic Mushrooms: The mushrooms that contain psilocybin are called magic mushrooms.

Malodour: Bad smell produced during composting.

Micron: A metric unit of measure equal to one one-thousandth (1/1000) of a Millimeter. Written as 0.001mm or 1μ or sometimes 1 μm.

Mycelium: The filamentous vegetative portion of a fungus, specifically excluding the fruiting structure or reproductive phase of the life cycle. The mycelium may be invisible or conspicuous. In some cases mycelial strands may join to form thick strands called rhizomorphs.

Mycophobia or Mushroom phobia: It is a Fear and Aversion to Mushrooms.

Mycorrhiza: A symbiotic relationship between a fungus and the roots of a plant.

Pallid: Pale, light in colour.

Parasitic: A fungus that grows at the expense of another organism, drawing nourishment from it. Example: *Armillaria mellea* (also called the honey mushroom or oak root rot fungus).

PCR: Polymerase Chain Reaction is a technique that allows an almost infinite amplification of a segment of DNA for which a primer (short piece of DNA) is available.

Phenolic: Having an odor of the chemical phenol.

Pileus (or Cap): The cap of a mushroom.

Polysaccharides: A large organic molecules consisting of many units of a simple sugar.

Psilocybin: It is a psychedelic drug with effects similar to that of LSD (Lysergic acid diethylamide).

Punctate: Having minute scales or points on the surface.

Putrescent: Tending to decay rapidly.

Reticulate: Usually referring to a fishnet or crosshatch pattern. The pattern may be on the surface of a mushroom as in *Bolbitius reticulatus,* the apex of the stipe as in *Boletus edulis,* or in the pattern of the ornamentation on some species of *Russula* or *Lactarius.*

Rhizomorphs: When fungal mycelium strands join to farm a thick strand, it is called rhizomorph (s).

Saprophytes: Fungi that receive nourishment from dead organic material.

Scabrous: Having conspicuous scales on the surface as in *Leccinum scabrum,* a common species introduced into California with the planting on non-native Birch trees.

Scrobiculate: Rounded depressions on the stipe of some species of mushrooms, as in *Lactarius scrobiculatus.*

Sinuate: Referring to a type of gill attachment, specifically gills that are notched at their point of attachment to the stipe.

SMS: The term SMS (Spent Mushroom Substrate) is used for used compost from mushroom cultivation. This spent compost is used for raising organic matter in the soil for field crops.

Spawn: Fungal propagule used for propagation of new mushroom crop (Latin *expandere* = to spread out). In other words, it is equivalent to seed of a plant (having no embryo) used for propagating mushroom.

Spawning: Spawning is the mushroom culture equivalent of planting seeds for a field crop.

Sphaerocyst: Round swollen cells usually formed in clusters, characteristically found in the Russulaceae. Sphaerocysts make the flesh of Russulas brittle.

Spore: Reproductive cell found in fungi. It is similar to seeds in higher plants, but devoid of embryo.

Sporocarp: The fungal reproductive structure, *e.g.* a mushroom that produces spores.

Sterilization: The elimination of organisms from containers by means of heat, chemicals, filters, etc.

Stipe (or Stem): The technical name for a mushroom stem or stalk. The stipe supports the pileus (cap) in the agarics (gilled mushrooms).

Stipitate: Having a stipe or stem.

Taxonomy: The classification of organisms to show relationships to other organisms.

Tomentose: Having very minute fine hairs on the surface.

Truncate: Having a flattened or chopped off end like the end of a baseball bat.

Tumour: An uncontrolled overgrowth of tissue (s).

Umbilicate: Having a small depression, *e.g.* as in a belly button.

Umbonate: Referring usually to the raised nipple-like structure at the center of some mushroom caps.

Viscid: Slimy, sticky, viscous surface of the pileus.

Volva: A sac-like structure formed at the base of a stipe, as it is found in *Amanita* species. It looks like a cup at the base of the stipe.

Zonate: Having a zoned appearance, usually referring to a mushroom cap that has concentric colour bands that give it a zoned appearance. Common in *Lactarius*.

APPENDIX

Appendix 1: Important Mushroom Journals

	Name of Journal	Published by
1.	Journal of Indian Mushroom Growers Association	Mushroom Research Laboratory, Chambaghat, Solan (Himachal Pradesh)
2.	Mushroom Research	Mushroom Society of India, National Centre of Mushroom Research and Training, Chambaghat, Solan
3.	Mushroom Science and The Mushroom Journal	Mushroom Growers Association, Agriculture House, Knight Bridge, SW IX, 7NJ, London, England
4.	Mushroom News	American Mushroom Institute, 907 E, Baltimore Pike Kennet Square, 1A 19348, USA
5.	Mushroom Journal for Tropics	Department of Botany, Chinese University of Hong Kong, New Territores, Hong Kong
6.	The Mycophile	The North American Mycological Association (NAMA) 336 Lenox Ave. Oakland, CA 94610-4675

Appendix 2: Important Books/Manuals dealing in Mushroom Science

1. Chang, S.T. and Hayes, W.A. The Biology and Cultivation of Edible Mushrooms. Academic Press Limited, London
2. Chang, S.T. and Miles, P.G. Edible Mushrooms and Their Cultivation. CRC Press, Boca-Raton, USA
3. Garcha, H.S. Manual of Mushroom Growing. PAU Publication, Ludhiana
4. Hobbs, Christopher, and Michael Miovic (ed.). Medicinal Mushrooms: An Exploration of Tradition, Healing and Culture. Third Edition. Botanical Press, Santa Cruz, CA.
5. Kannaiyan, S. and Ramaswamy, K. Handbook of Edible Mushrooms. Today and Tomorrow's Printers and Publishers, New Delhi
6. Kapoor, J.N. Mushroom Cultivation. ICAR Publication, New Delhi
7. Bahl, Nita. Hand Book on Mushrooms. Oxford and IBH, New Delhi, India
8. Oei, Peter. Manual on Mushroom Cultivation: Techniques, Species and Opportunities for Commercial Application in Developing Countries. TOOL Publications, Amsterdam, The Netherlands.
9. Purkayastha, R.P. and Chandra, A. Manual of Indian Edible Mushrooms. Today and Tomorrow's Printers and Publishers, New Delhi
10. Singh, H. Mushrooms-The of Cultivation. Sterning Publishers Pvt. Ltd., New Delhi, India
11. Singh, R.P., Kumar, A and Kumar, P. Mushroom Production Techniques. Jain Offset Printers, New Delhi
12. Stamets, Paul, and J.S. Chilton. The Mushroom Cultivator. Agarikon Press, Olympia, WA.

13. Stamets, Paul. Growing Gourmet and Medicinal Mushrooms. Ten Speed Press. Berkeley, CA.
14. Tewari, S.C. and Kapoor, P. Mushroom Cultivation: An Economic Analysis. Oxford and IBH, New Delhi, India

Appendix 3: Important Bulletins dealing in Mushroom Science

1. Chahal, D.S. Mushroom Growing for beginners. *Punjab Agricultural University,* Ludhiana-141 004 (Punjab)
2. Gupta, Y. and Sharma, S.R. Mushroom Spawn Production. *National Centre for Mushroom Research and Training,* Solan-173 213 (H.P.)
3. Munjal, R.L. Mushroom for Every Home. *Indian Agricultural Research Institute,* New Delhi
4. Saxena, S. and Rai, R.D. Post Harvest Technology of Mushrooms. *National Centre for Mushroom Research and Training,* Solan-173 213 (H.P.)
5. Singh, R.P. Successful Mushroom Production. *G.B. Pant University of Agriculture and Technology,* Udham Singh Nagar, Uttranchal
6. Sohi, H.S. Mushroom Cultivation. *Indian Institute of Horticultural Research,* Hessaraghat-560 089, Bangalore
7. Vijay, B. and Gupta, Y. Cultivation of White Button mushroom (*Agaricus bisporus*). *National Centre for Mushroom Research and Training,* Solan-173 213 (H.P.)

Appendix 4: Mushroom Species and their Nicknames

Mushroom Species	Nickname (s)
Agaricus arvensis	Horse mushroom
Agaricus bisporus	Button mushroom
Agaricus portobello	Portobello or Italian Agaricus
Agaricus. Brunnescens	White Button
Amanita caesarea	Caesar's Mushroom
Amanita muscaria	Fly Agaric
Amanita pantherina	Panthercap
Amanita phalloides	Death Cap
Amanita verna	Fool's Mushroom
Amanita virosa	Destroying Angel
Armillariella mellea	Honey Mushroom
Auricularis species	Ear mushroom
Boletus piperatus	Pepper bolete
Cantharellus species	Chanterelles
Chlorophyllum molybdites	Green Gill mushroom
Coprinus species	Shaggy Mane mushroom
Cortinarius orellanus	The Sorrel Webcap mushroom
Entoloma lividum	Gray Pink Gill
Flammulina species	Winter mushroom
Flammulina velutipes	Enoki or Enokitake or winter mushrooms
Galerina autumnalis	Autumn Skullcap
Ganoderma lucidum	Reishi mushroom
Ganoderma species	Reishi mushroom
Grifola frondosa	Maitake mushroom

Gymnopilus spectabilis	Showy Flamecap or Big Laughing Mushroom
Gyromitra esculenta	False Morel
Gyromitra gigas	False Morel
Hericium erinaceus	Lions Mane
Hericium species	Lion's Head or Pom Pom mushroom
Laetiporus sulphureus	Sulfur shelf fungus
Lentinus species	Shiitake mushroom
Leucocoprinus procera	Parasol Mushroom
Leucocoprinus rachodes	Shaggy Parasol
Omphalotus illudens	Jack O'Lantern
Omphalotus olearius	Jack O'Lantern Mushroom
Paxillus involutus	Naked Brimcap
Pholiota nameko	Nameko mushroom
Pholiota species	Nameko mushroom
Pleaurotus djamor	Pink Oyster
Pleurotus ostreatus	Pearl Oyster
Pleurotus pulmonarius	Grey Oyster
Pleurotus species	Oyster mushroom
Polyporus frondosus	Hen of the Woods or Maitake mushroom
Polyporus sulphureus	Chicken of the Woods
Russula emetica	Sickener
Russula virescens	Green Bitter Gill
Tremella species	White Jelly mushroom
Tricholoma myomyces	Waxygill Cavalier
Tricholoma pardinum	Tigertop

Tricholoma virgatum	Silver Streaks
Tricholomopsis platyphylla	Broad Gill
Verpa bohemica	Early False Morel
Volvariella species	Straw mushroom
Volvariella violaceae	Patty Straw Mushroom
Hydnum spandum	Doe skin mushroom

Appendix 5: Important Spawn suppliers

1. Agro Products, E-18, Green Park Extension, New Delhi
2. Centre for Tropical Mushroom Research and Training, Department of Plant Pathology, Orissa University of Agriculture and Technology, Bubhnewar, Orissa
3. Department of Horticulture, Public Garden, Hyderabad, Andhra Pradesh
4. Diamond Mushroom Products, Ambatinagar, Mangalgiri Mandal, Andhra Pradesh
5. Division of Plant Pathology, Indian Agricultural Research Institute, New Delhi
6. Flex Food Limited, Lal Chhappar Industrial Area, Dehradun, Uttranchal
7. Golden Food Export, Lanugs Garden Road, Pudanpur, Chennai, Tamil Nadu
8. Himalayan Spawn Laboratory, Kandaghat, Solan, Himachal Pradesh
9. Mushroom Research Centre, G.B. Pant University of Agriculture and Technology, Udham Singh Nagar, Uttranchal
10. Mushroom Research Laboratory, Division of Plant Pathology, Indian Institute of Horticultural Research, Hessarghatta, Bangalore, Karnataka

11. National Centre on Mushroom Research and Training, Solan, Himachal Pradesh
12. Snow view Mushroom Laboratory and Training Centre, Gandhi Ashram, Narela, New Delhi
13. Spawn Laboratory, Department of Horticulture, Jalandhar
14. Spawn Production Laboratory, Lalmandi, Srinagar
15. Spawn Production Laboratory, Talab Tillo, Jammu, Jammu and Kashmir
16. Swadesh Mushroom Farm, B-28, G.T. Karnal Road, Industrial Area, New Delhi

Appendix 6: Important International Spawn Suppliers

1. Amycel Spawn, P.O. Box 560, Avondale, PA 19311
2. Choice Edibles, 84 Riverside Park Road, Carlotta, CA 95528
3. Crop King, 5050, Greenwich Road, Seville, OH 44273-9413
4. Fungi Perfecti, P.O. Box 7634, Olympia, WA 98507
5. Gourmet Mushroom Products, P. O. Box 515 IP, Graton, CA 95444
6. L.F. Lambert Spawn Co., 1507 Valley Rd., Coatesville, PA 19320
7. Sylvan Spawn Laboratory, West Hills, Industrial Park, Kittanning, PA 16201

Appendix 7: Important Equipment Suppliers

1. D.V. Agro Industries Private Limited, B-325, New Friends Colony, New Delhi-110 085
2. Master Mechanical Works, 'Pushpanjali' S.V. Road, Santacruze (W), Bombay-400 047

3. Techno-equipments, 31, Parekh Road, Bombay-400 004
4. Teg's Trymsater, F-17, Hauz Khas Enclave, New Delhi-110 016
5. Temp Master Enterprises, 15-16/7, Community Centre, Naraina Industrial Area Phase I, New Delhi-110 028
6. Unicorn Implements & Manufacturing Corporation, 113 Highway, 24 Commerce, Texas 75429 USA
7. Indvent engineers Private Limited, C-12, Amar Colony Market, Lajpat Nagar, New Delhi-110 024
8. Urjex Industries, S-26, Industrial Estate, Partpur, Meerut-250 102

Appendix 8: Exporters of Mushrooms

1. Ankur internation, 2147, Sector-35-C, Chandigarh-160 036
2. Asiatic Agencies, 4850, Ansari Road, Darya Ganj, New Delhi-110 002
3. Calcutta Cellulose Private Limited, 4-C, Meherali Mondal Street, Calcutta-700 027
4. Dhanuka Exporters Private Limited, M-279, greater Kailash, Part II, New Delhi-110 048
5. K.V. Exports, 13, jeevan Prag, Prabhat Colony, Road No. 2, Santacruz (E), Mumbai-400 055
6. Rajshree Exports, 76-79A, Industrial Area, Govindpura, Bhopal-462 023
7. Venturex Processed Food Private Limited, 1002, 10th floor, Babukhan Estate, Basheerbagh, Hyderabad- 500 001

Appendix 9: Other uses of mushrooms

1. The fruit body of *Fomentarius* and *Ganoderma appalatum* is used for producing Suede like materials.
1. Bottle corks are made from *polyporus squasmosus*.
2. *Polyporus formentarius* also known as timber mushroom is used for cigar light.
3. *Polyporus nigricans* is used for preparing snuffing agents.
4. *Polyporus bispidus* is used for making dye for colouring silk, cotton and wool.
5. *Daedalea quercika* is used as a comb to rub the body of horses.
6. *Coprinus comatus* is used for making black ink.

Appendix 10: Various formulations available for cultivation

1. Formulations for Natural Compost

1.	Horse manure	1000 kg
2.	Wheat straw	350 kg
3.	Urea	3 kg
4.	Gypsum	30-40 kg

Developed from IARI, New Delhi

1.	Horse manure	1000 kg
2.	Wheat straw	500 kg
3.	Chicken manure	300 kg
4.	Brewer's grain	60 kg
5.	Gypsum	30 kg

Developed from IARI, New Delhi

2. Formulations for Synthetic compost

i) Formulae given by Indian Institute of Horticultural Research, Hessarghatta, Bangalore

Formula 1

1.	Paddy straw	150 kg
2.	Maize stover	150 kg
3.	Ammonium sulphate	9 kg
4.	Superphosphate	9 kg
5.	Urea	4 kg
6.	Rice bran	50 kg
7.	Gypsum	12 kg
8.	Calcium carbonate	10 kg
9.	Cotton seed meal	5 kg

Formula 2

1.	Wheat straw	300 kg
2.	Calcium Ammonium Nitrate	9 kg
3.	Superphosphate	9 kg
4.	Urea	4 kg
5.	Wheat bran	30 kg
6.	Gypsum	12 kg
7.	Calcium carbonate	10 kg

Formula 3

1.	Wheat straw	300 kg
2.	Chicken manure	120 kg
3.	Rice bran	20.6 kg
4.	Brewer's grain	22 kg
5.	Urea	6 kg
6.	Cotton seed meal	15 kg

ii) Formulation given by NCMRT, Solan

Formulae 1

No.	Ingredient	Quantity
1.	Wheat straw	300 kg
2.	Wheat bran	15 kg
3.	Chicken manure	125 kg
4.	Urea	5.5 kg
5.	Gypsum	20 kg
6.	pesticide	

Formulae 2

No.	Ingredient	Quantity
1.	Wheat straw and paddy straw (1:1)	300 kg
2.	Calcium Ammonium Nitrate	9 kg
3.	Urea	4 kg
4.	Wheat bran	25 kg
5.	Gypsum	20 kg
6.	Pesticide	

Formulae 3

No.	Ingredient	Quantity
1.	Wheat Straw	1000 kg
2.	Calcium ammonium Nitrate (CAN)	30 kg
3.	Super Phosphate Single (SSP)	25 kg
4.	Urea	12 kg
5.	Sulphate of Potash	10 kg
6.	Wheat Bran	100 kg
7.	Molasses	16.5 Lt
8.	Nematicide	

3. Formulations for Semi-Synthetic compost

Formula 1

1.	Horse Manure	200 kg
2.	Wheat Straw	100 kg
3.	Poultry Manure	50 kg
4.	Wheat Bran	15 kg
5.	Urea	3 kg
6.	Gypsum	10 kg

Formula 2

1.	Paddy Straw	33 kg
2.	Poultry Manure	10 kg
3.	Wheat Bran	1.5 kg
4.	Gypsum	1 kg

REFERENCES

Ajith T.A., Janardhanan, K.K. 2001. Antioxidant and anti-inflammatory activities of methanol extract of *Phellinus rimosus* (Berk) *Pilat. Indian J Exp Biol.* 39(11): 1166-9.

Atkins, Fred C. 1974. Guide to Mushroom Growing. Faber and Faber Ltd., 3 Queen Squares, London.

Bano Z, Rajarathnam S. 1988. Pleurotus mushrooms. Part II. Chemical composition, nutritional value, post-harvest physiology, preservation, and role as human food. *Crit Rev Food Sci Nutr.* 27(2): 87-158.

Benjamin, D.R., 1995. Mushrooms: Poisons and Panaceas. A Handbook for Naturalists, Mycologists and Physicians, New York: Freeman and Co.

Blackwell, W.H., Poisonous and Medicinal Plants. Englewood Cliffs: Prentice Hall, 1990.

Blum, H. 1977. The Mushroom Industry in Ontario. Economic Branch, Ontario Ministry of Agriculture and Food, Toronto, Ontario.

Borchers, A.T., Stern, J.S., Hackman, R.M., Keen, C.L., Gershwin, M.E. 1999. Mushrooms, tumors, and immunity. *Proc Soc Exp Biol Med.* 221(4): 281-93.

Chang, S.T. and W. A. Hayes. 1978. The Biology and Cultivation of Edible Mushrooms. Academic Press, New York.

Chang, S.T. 1999a. Global impact of edible and medicinal mushrooms on human welfare in the 21st Century: non-green revolution. *International Journal of Medicinal Mushrooms* 1, 1-7.

Chang, S.T. 1999b. World production of cultivated edible and medicinal mushrooms in 1997 with emphasis on *Lentinus edodes* (Berk.) Sing. in China. *International Journal of Medicinal Mushrooms* 1, 291-300.

Chang, S.T. and Miles, P.G. 1989. *Edible mushrooms and their Cultivations*. CRC Press Inc., Boca Raton, Florida.

Chu K.K., Ho, S.S., Chow, A.H. 2002. *Coriolus versicolor*: a medicinal mushroom with promising immunotherapeutic values. *J Clin Pharmacol*. 42(9): 976-84.

Dang, R.L. and Singh, R.P. 1978. Preservation of Mushrooms. *Indian Mush. Sci.* 1: 215-224.

Dutta, R. 2005. Mushroom Cultivation. **In:** Agriculture Refresher Dutta, R. (edn). Jain Brothers, New Delhi.

Dutta, R. 2004. Technology for Cultivation of Button Mushroom. *Archana*. 40**:** 60-62.

Fullerton, S.A., Samadi, A.A., Tortorelis, D.G., Choudhury, M.S., Mallouh, C., Tazaki, H., Konno, S. 2000. Induction of apoptosis in human prostatic cancer cells with beta-glucan (Maitake mushroom polysaccharide). *Mol Urol*. 4(1): 7-13.

Gupta, G.K., Bajaj, B.S. and Suryanarayan, D. 1970. Studies on the cultivation of Paddy straw mushroom (*Volvariella volvacea, V. diplacea*). *Indian Phytopath.* 23: 615-620.

Hamlyn, P.F. (1989). Cultivation of Edible Mushrooms on Cotton Waste. *The Mycologist* 3(4): 171-173.

Hayes, W.A. and Haddad, N. 1976. The food value of the cultivated mushrooms and their importance to the mushroom industry. *The Mushroom J.* 40: 104-110.

Ho, M.S. 1989. A new technology "Pastic Bag cultivation method" for shiitake mushroom. *Mush. Sci.* 12(2): 303-309.

Jandiak, C.L. 1995. Economics of Mushroom Cultivation. **In:** Advances in Horticulture Vol.13 Eds. Chadha, K.L. and Sharma, S.R. Malhotra Publishing House. New Delhi.

Jandiak, C.L. and Kapoor, J.N. 1974. Studies on the cultivation of *Pleurotus sajor-caju* (Fr) Singer. *Mush Sci.* IX: 667-672.

Ji, Lelley. 1999. From edible to useful mushrooms—an attempt for the new economical assessment of large fungi. *Acta Microbiol Immunol Hung.* 46(2-3): 205-13.

Kubo, K., Nanba, H. 1996. The effect of maitake mushrooms on liver and serum lipids. *Altern Ther Health Med.* 2(5): 62-6.

Kuhn, C., Swartzwelder, S. and Wilson, W. Buzzed. 1998. The Straight Facts About The Most Used and Abused Drugs from Alcohol to Ecstasy, New York: W.W. Norton and Company, 1998.

Lambert, L. F. 1958. Practical and Scientific Mushroom Culture. L. F. Lambert, Inc. Coatesville, PA 19230.

Leatham, G.F. (1982). Cultivation of shiitake, the Japanese forest mushroom, on logs: a potential industry for the United States. *Forest Products Journal*, 32(8), 29-35.

Lintzel, W. 1941. The nutritional value of edible mushroom protein. *Biochem. Acta.* 308: 413-419.

Matsuzawa T, Saitoh H, Sano M, Tomita I, Ohkawa M, Ikekawa T. 1998. Studies on antioxidant effects of Hypsizigus marmoreus. II. Effects of Hypsizigus marmoreus for antioxidant activities of tumor-bearing mice. *Yakugaku Zasshi.* 118(10): 476-81.

Mattila, P., Konko, K., Eurola, M., Pihlava, J.M., Astola, J., Vahteristo, L., Hietaniemi, V., Kumpulainen, J., Valtonen, M., Piironen, V. 2001. Contents of vitamins, mineral elements, and some phenolic compounds in cultivated mushrooms. *J Agric Food Chem.* 49(5): 2343-8.

Mau J.L., Chao, G.R., Wu, K.T. 2001. Antioxidant properties of methanolic extracts from several ear mushrooms. *J Agric Food Chem.* 49(11): 5461-7.

Mau, J.L., Lin, H.C., Chen, C.C. 2002. Antioxidant properties of several medicinal mushrooms. *J Agric Food Chem.* 50(21): 6072-7.

Mutanen, M. 1986. Bioavailability of selenium in mushrooms, Boletus edulis, to young women. *Int J Vitam Nutr Res.* 56(3): 297-301.

Padwick, G.W. 1991. Mushroom Cultivation in India. *Indian Farming.* 11: 363-366.

Parslew, R., Jones, K.T., Rhodes, J.M., Sharpe, G.R. 1999. The antiproliferative effect of lectin from the edible mushroom (Agaricus bisporus) on human keratinocytes: preliminary studies on its use in psoriasis. *Br J Dermatol.* 140(1): 56-60.

Royse, D.J. 1985. Shiitake mushrooms - consumption, production and cultivation. *Interdisciplinary Science Reviews,* 10(4), 329-335.

Shouji, N., Takada, K., Fukushima, K., Hirasawa, M. 2000. Anticaries effect of a component from shiitake (an edible mushroom). *Caries Res.* 34(1): 94-8.

Spoerke, D.G. and Hall, A.H. 1990. Plants and mushrooms of abuse. *Emerg. Med. Clin. North Amer.,* 8: 579-593.

Stijve T. 1977. Selenium content of mushrooms. *Z Lebensm Unters Forsch.* 164(3): 201-3.

Swayne, J.B. 1950. Handbook of Mushroom Culture, Kennett Square, PA 19348.

Vedder, P. J. C. 1978. Modern Mushroom Growing. Pitman Press, Bath, G. B. Distributed in U.S.A. by S.A.S., Inc., RFD 1, Box 80 A, Madisonville, TX 77864.

Wasser, S.P., Weis, A.L. 1999. Therapeutic effects of substances occurring in higher Basidiomycetes mushrooms: a modern perspective. *Crit Rev Immunol.* 19(1): 65-96.

Wasser, S.P. 2002. Medicinal mushrooms as a source of antitumor and immunomodulating polysaccharides. *Appl Microbiol Biotechnol.* 60(3): 258-74.